Optoelectronics of Solar Cells

Also Available from SPIE Press Exclusively Published by New Age International Pvt. Ltd.

Fourier-Transform Spectroscopy Instrumentation Engineering
Vidi Saptari
ISBN: 978-81-224-2823-0, Original Edition Price: $ 48.00, Indian Edition Price: Rs. 395.00

Tissue Optics : Light Scattering Methods and *Instruments for Medical Diagnosis,* 2nd Ed.
Valery Tuchin
ISBN: 978-81-224-2829-2, Original Edition Price: $ 99.00, Indian Edition Price: Rs. 995.00

Local Approximation Techniques in Signal and Image Processing
Vladimir Katkovnik, Karen Egiazarian, Jaakko Astola
ISBN: 978-81-224-2830-8, Original Edition Price: $ 99.00, Indian Edition Price: Rs. 695.00

Logic-based Nonlinear Image Processing
Stephen Marshall
ISBN: 978-81-224-2825-4, Original Edition Price: $ 51.00, Indian Edition Price: Rs. 395.00

Introduction to the Optical Transfer Function
Charles S. Williams, Orville A. Becklund
ISBN: 978-81-224-2831-5, Original Edition Price: $ 78.00, Indian Edition Price: Rs. 695.00

Digital and Analog Fiber Optic Communications for CATV and FTTx Applications
Avigdor Brillant
ISBN: 978-81-224-2832-2, Original Edition Price: $ 115.00, Indian Edition Price: Rs. 999.00

Infrared Fibers and Their Applications
James A. Harrington
ISBN: 978-81-224-2827-8, Original Edition Price: $ 78.00, Indian Edition Price: Rs. 495.00

Optoelectronics of Solar Cells
Greg P. Smestad
ISBN: 978-81-224-2826-1, Original Edition Price: $ 46.00, Indian Edition Price: Rs. 395.00

MOEMS: Micro-Opto-Electro-Mechanical Systems
Manouchehr E. Motamedi
ISBN: 978-81-224-2833-9, Original Edition Price: $ 113.00, Indian Edition Price: Rs. 795.00

Introduction to Image Stabilization
Scott W. Teare, Sergio R. Restaino
ISBN: 978-81-224-2822-3, Original Edition Price: $ 51.00, Indian Edition Price: Rs. 395.00

Fundamentals of Polarimetric Remote Sensing
John R. Schott
ISBN: 978-81-224-2834-6, Original Edition Price: $ 68.00, Indian Edition Price: Rs. 395.00

Fundamentals of Infrared Detector Materials
Michael A. Kinch
ISBN: 978-81-224-2824-7, Original Edition Price: $ 61.00, Indian Edition Price: Rs. 395.00

Optoelectronics of Solar Cells

Greg P. Smestad

SPIE
PRESS
Bellingham, Washington USA

NEW AGE

First Indian Sub-Continent Edition : 2010
by New Age International (P) Ltd., Publishers

Branches:

- 36, Malikarjuna Temple Street, Opp. ICWA, Basavanagudi, **Bangalore**. ✆ (080) 26677815
- 26, Damodaran Street, T. Nagar, **Chennai**. ✆ (044) 24353401
- CC-39/1016, Carrier Station Road, Ernakulam South, **Cochin.** ✆ (0484) 4051303
- Hemsen Complex, Mohd. Shah Road, Pal n Bazar, Near Starline Hotel, **Guwahati**. ✆ (0361) 2543669
- No. 105, 1st Floor, Madhiray Kaveri Tower, 3-2-19, Azam Jahi Road, Nimboliadda, **Hyderabad**. ✆ (040) 24652456
- RDB Chambers (Formerly Lotus Cinema) 106A,1st Floor, S.N. Banerjee Road, **Kolkata**. ✆ (033) 22275247
- 16-A, Jopling Road, **Lucknow**. ✆ (0522) 2209578
- 142C, Victor House, Ground Floor, N.M. Joshi Marg, Lower Parel, **Mumbai**. ✆ (022) 24927869
- 22, Golden House, Daryaganj, **New Delhi**. ✆ (011) 23262370, 23262368

Printed in India at V. K. Printers, Delhi.
C-09-12-4223

ISBN: 978-81-224-2826-1
Original Edition Price: $ 46.00
Indian Edition Price: Rs. 395.00

PUBLISHING FOR ONE WORLD
NEW AGE INTERNATIONAL (P) LIMITED, PUBLISHERS
4835/24, Ansari Road, Daryaganj, New Delhi-110002
Visit us at **www.newagepublishers.com**

A new scientific truth does not triumph by convincing its opponents, but rather because its opponents die, and a new generation grows up that is familiar with it.

Max Planck

There is one thing stronger than all the armies in the world: and that is an idea whose time has come.

Victor Hugo

Table of Contents

List of Symbols and Terms

Symbols and terms for solar cells come from many fields. These are the symbols most often used in practice.

A	Physical area
a	Acceptor (as subscript)
α	Absorption coefficient (i.e., in cm^{-1})
a(e)	Absorptivity. This can also be expressed as a function of the wavelength, $a(\lambda)$.
AM1.5	Air Mass 1.5. The standard solar spectrum and light output used for solar-cell measurement. It is equal to the reciprocal of the cosine of the angle of the sun.
C	Concentration ratio (optical gain). Also, the concentration of a liquid solution as used in the Beer-Lambert law.
CB	Conduction band
c	Velocity of light
C-V	Cyclic voltammetry. An electrochemical characterization technique.
D	Diffusion coefficient
Dye	A chemical pigment molecule or compound that absorbs visible light
ε (e)	Quantum emissivity used in the Planck equation [$\varepsilon(e) = a(e)$ via Kirchhoff's law]
ε	Molar extinction coefficient used in the Beer-Lambert law
e	Energy (e.g., photon or phonon energy)
exp	Natural exponential (i.e., base e)
E_g or e_g	Bandgap energy
E_{fn}	Fermi level of electrons

F_{fp}	Fermi level for holes
E_{redox}	Redox couple midpoint potential energy
E	Energy
E_f	Fermi energy
E_p	Phonon energy
η	Solar conversion efficiency. The ratio of power output to total power input.
η_c	Collection efficiency
η_q	Quantum efficiency (also called IPCE). The probability that an electron injected into a material will be collected at the electrical contact.
φ	Azimuthal angle used in a spherical coordinate system
f	Dilution factor used in Planck equation. A geometrical factor.
FF	Fill Factor. Ratio of maximum power produced to product of open circuit voltage and short circuit current.
FTIR	Fourier transform infrared. A type of spectroscopy.
Φ	Luminescence efficiency (i.e., fluorescence yield)
Γ	Photon flux density (number per unit time per unit area)
Γ_{0R}	Radiative recombination flux density
γ	Diode quality, or ideality factor, in the diode equation
G	Generation rate of charge carriers via the absorption of light
h	Planck's constant
HOMO	Highest occupied molecular orbital. The lower energy level of a molecule.
Hole	Site in a crystal where an electron is missing. It is thus positively charged, and sometimes mobile.
Injection	The transfer of an electron or hole into a semiconductor or solid (a process in dye sensitization).
I	Current (e.g., mA)
I_{SC}	Short circuit current
I_0	Blackbody current (Ideally, $q\Gamma_0$). The reverse saturation current.
IPCE	Incident photocurrent efficiency. Also given the symbol η_q.

I-V	Current-voltage (compare with J-V)
J	Current density. Current per unit area (i.e., mA/cm^2)
k	Boltzmann's constant
κ	Extinction coefficient
K	Ratio of nonradiative to radiative losses in a solar cell material
L	Diffusion length for a charge carrier used in the transport equation. Also used as the radiance in the Planck equation in optics.
L_x	Radiance of the light of origin (x). Other names include brightness and luminance.
L_{0R}	Radiance of light emitter at ambient temperature
L_S	Radiance of source
ln	Natural logarithm
λ	Wavelength of light (e.g., in μm or nm).
LUMO	Lowest unoccupied molecular orbital. The higher energy level of a molecule.
mega-	The prefix of a unit meaning 1,000,000, or 10^6 of that unit (e.g., Mohm). A Megawatt is a million watts or a thousand kilowatts.
mediator	In a photoelectrochemical cell, the mediator is the substance that maintains (and protects) the neutral electric charge on another substance by becoming oxidized or reduced. In the dye-sensitized solar cell, it is the catalyst present in the electrolyte between the TiO_2 coated glass and counter electrode.
micro-	A prefix meaning 10^{-6} of a unit (e.g., μA).
milli-	A prefix meaning 10^{-3} of a unit (e.g., mA).
Mohm	Mega ohm (10^6 ohms). A measure of electrical resistance.
μ	Chemical potential. Also, in semiconductor physics, the mobility of a charge carrier.
μ_{max}	Maximum chemical potential
n^*	Maximum concentration of photoexcited electrons
nano-	A prefix meaning 10^{-9} of a unit
n_p	Electron concentration or electron density on the P-type side of the solar converter. Electron concentration on the "pump" side of the solar converter.

n_n	Electron concentration on the N-type (or acceptor) side of the converter.
NHE	Normal hydrogen electrode reference potential
N_A	Dopant concentration for acceptor atoms (i.e., cm^{-3})
N_D	Dopant concentration for donor atoms (i.e., cm^{-3})
n_i	Intrinsic electron concentration
n_0	Electron concentration in the dark
n_{p0}	Electron concentration in the dark on the p-type side of the solar cell
n	Optical index of refraction. As a subscript, an electron. Also, electron concentration.
n_c	Complex index of refraction consisting of both real and imaginary portions
n_p	Concentration of electrons on pump or P-type side of the solar cell
p	As a subscript, pump or pertaining to holes. In P-N junction solar cells, the p-doped base layer
p^*	Maximum concentration of photoexcited holes
p_p	Concentration of holes on the P-type side of the cell
p_0	Hole concentration in the dark (e.g., at equilibrium)
P_{abs}	Net absorbed radiant power
PEC	Photo-electrochemical (solar) cell
PV	Photovoltaic cell. A device that converts radiant energy (photons or light) into electricity. A solar cell.
q	Charge on the electron
Ref	Optical reflectivity of a surface
RC	Recombination centers
R_S	Series resistance
R_{sh}	Shunt resistance
r	Radiative or pertaining to radiative processes (as a subscript)
Ru Dye	cis- $(SCN)_2$ Bis(2, 2' bipyridyl - 4,4' - dicarboxylate) ruthenium(II) Charge-Transfer Sensitizer

$\dot{S}_{tot}$	Total entropy generation rate
SEM	Scanning electron microscope or microscopy
ΔS	Change in entropy per mol
t	Thickness
T(e)	Transmission coefficient as a function of photon energy
T_1	Transmission coefficient for a single pass through the absorber
T_S	Source or solar temperature
T_0	Ambient temperature (i.e., 300 K)
TRMC	Time-resolved microwave conductivity
TiO_2	Titanium dioxide, titania. An inert mineral used in pigments and industrial applications
τ	Lifetime of the excited state
UV	Ultraviolet (light in the range from 200–400 nm)
θ	Angle with respect to the surface normal
μ	Charge carrier mobility. Also used as the symbol for chemical potential.
U	Recombination rate
V	Voltage. A measure of difference in electrical potential between two electrodes or points (in volts).
V_{OC}	Open circuit voltage
VB	Valence band
Vis	Visible light in the range from 400–700 nm
W	Work (e.g., in units of Joules or eV)
W_{rev}	Reversible work
x_p	Location of P-N junction edge
x	General distance
x_g	e_g/kT_0
y	$(x-x_p)$

Units and Useful Numerical Quantities

Electron Charge:	$q = 1.6022 \times 10^{-19}$ C
Boltzmann's constant:	$k = 1.38066 \times 10^{-23}$ J/K $= 8.617 \times 10^{-5}$ eV/K
Plank's constant:	$h = 6.6261 \times 10^{-34}$ J • s $= 4.136 \times 10^{-15}$ eV• s
Speed of light in vacuum:	$c = 2.99792 \times 10^{8}$ m/s
Thermal Voltage at 300 K:	$kT/q = 0.0259$ V
Wavelength of 1 eV photon:	1.23977 μm
Stefan–Boltzman constant:	5.6703×10^{-8} W K^{-4} m^{-2}
Avogadro's number:	6.0225×10^{26} per mol
Pi:	$\pi = 3.141592654...$
	$2\pi/ h^3c^2 = 9.883 \times 10^{26}$ eV^{-3} s^{-1} m^{-2}
e:	$e = 2.718282...$

Length:

1 meter (m)	3.28 feet (ft)
1 kilometer (km)	0.621 miles

Weight:

1 kg (kg)	2.2 pounds (lb)
1 metric ton	1,000 kg = 1.1 short tons

Area:

1 m^2	10.75 ft^2
1 km^2	0.386 $mile^2$ = 100 hectares
1 hectare (ha)	2.47 acres = 10,000 m^2

Temperature: T(Kelvin) = T(Celsius) + 273.15

Energy:

1kWh	1000 Wh $= 3.6 \times 10^6$ J = 860.4 kcal = 3413 Btu
1 eV	1.6022×10^{-19} J

Power:

1 MW	1000 kW $= 10^6$ W $= 10^6$ J/s
1 kW	0.239 kcal/s = 1.341 hp = 3413 Btu/h

Solar power incident at Earth's atmosphere = 173,000 TW $= 173{,}000 \times 10^{12}$ W

Solar power conversion by photosynthesis = 50 to 100 TW

Power used by humans ≈ 13 TW (2002)

Preface

With concerns about worldwide environmental security, global warming, and climate change due to emissions of CO_2 from the burning of fossil fuels, it is desirable to have a wide range of energy technologies in a nation's portfolio. These technologies can be used in domestic markets, or exported to other nations, helping them to "leapfrog" to a cleaner, and less carbon intensive, energy path. Far from being an altruistic act, these energy technologies are lucrative businesses that will grow stronger in the global economy of the 21st century. According to U.S. DOE EIA, NREL U.S. PV Industry Technology Roadmap 1999 Workshop and Strategies Unlimited, photovoltaics (or PV) is a billion dollar a year industry and is expected to grow at a rate of 15–20% per year over the next few decades. Solar cells have already proven themselves a viable option as a nonpolluting renewable energy source in many applications. It is advantageous to optical engineers to have at least a basic knowledge of how these devices function, and of the important parameters that control their operation. This text is designed to be an overview for those in the fields of optics and optical engineering, as well as those who are interested in energy policy, economics, and the requirements for efficient photo-to-electric energy conversion.

Greg P. Smestad
April 2002

Optoelectronics of Solar Cells

1

Introduction to Solar Cells

1.1 Philosophy of the Text

Many basic texts already exist that describe the materials science aspects of solar cells and solar photovoltaic (PV) modules. The nomenclature used to describe solar cells in these texts is often difficult for the novice to understand and relate to basic concepts in optics. This tutorial text utilizes many of these materials science and solid-state physics texts as references, but it takes a slightly different approach. The goal herein is a description of the basic function of solar cells from an optical perspective. The fundamental principles of photovoltaic solar converters are examined with emphasis on their optical properties, and the requirements for the production and manufacturing of efficient and cost-effective light converters.

Rather than describing all the numerous specific examples of solar cells that include silicon (Si), copper indium diselenide ($CuInSe_2$), gallium arsenide (GaAs), and cadmium telluride (CdTe), among others, the approach here is to give a broader view of what these devices actually have in common. In typical texts, the description of solar device performance is made in terms of Fermi statistics (e.g., of electrons). In these texts, doping and electron concentration profiles are described, as well as specific differences in solar cell geometries and methods of their construction discussed. In contrast, this text concentrates on describing solar cells using Boson statistics for photons. Although both approaches are outlined and presented, the emphasis is on the Planck equation rather than on the Fermi equation. In other words, we will view things from the standpoint of the photon rather than the electron. Hopefully, this distinction will allow the reader to understand the basics of solar cell-device design, as well as developments that will occur in the future. This text provides a background from which the reader may delve into other texts in order to gain a further practical understanding of solar cell-device structures. Throughout the text, general methods of solar cell testing and characterization are also outlined. The methodology in this text is by no means novel, and is a result of the work of many researchers who have shared the goal of understanding the fundamental limitations to conversion efficiency. These researchers agree that the performance of solar cells is determined by how a material absorbs, reflects, uses, and even emits light.

1.2 Renewable Energy and Photovoltaics Background

It is useful to give some background on renewable energy and photovoltaic terminology before proceeding to describe the solar cell. Fossil fuels, like coal, natural gas, and oil, provide power to our society and add to the input of solar energy we receive daily. This "fossil" fuel reserve is finite. It is estimated that only a few hundred years supply is available at our current rate of consumption and that rate is increasing as developing nations compete for their share of the global economy. This resource problem aside, the use of these stored products of ancient photosynthesis is not without a price. As the carbon-dioxide concentration in the atmosphere increases due to this fossil-fuel consumption and deforestation, we are inadvertently returning Earth to the state of high carbon-dioxide concentration found when the sun was much weaker. It is predicted by the United Nations Intergovernmental Panel on Climate Change that the warming of Earth over the next 50 years could have devastating effects on sea levels, agriculture, climate, immigration, and economic development. The benefits of diversifying our energy portfolio include those mentioned above, as well as economic benefits to both the United States and to the rest of the world due to the use of energy resources available locally.

One method of taking advantage of these benefits and weaning ourselves away from fossil fuels is to use renewable and alternative energy directly to produce fuels and electricity. Today, natural photosynthesis on the land and in the oceans produces eight times the current combined energy requirements of humanity. Converting solar energy at only 10% efficiency using 1% of Earth's land area would supply us with twice our current energy needs worldwide. Using current solar technologies, an area defined by a square 161 km (100 miles) on a side located in a sunny area could produce, in one year, the energy equivalent to that used annually in the entire United States. This area could be centrally located, or distributed on rooftops throughout an area where energy is needed. The challenge then becomes harvesting this energy in an economically efficient way.

Over the past several decades, researchers have advanced solar technologies and learned how to use materials to create solar converters rivaling those of nature. One renewable energy technology uses photovoltaic (PV) solar cells, which convert incoming solar radiation directly into electricity. PV modules are large-area solid-state semiconductor devices that convert solar energy directly into electricity.

The history of the solar cell starts in the late 19th century with the principles of photography. It was discovered that silver chloride (and other silver halides) respond to light. The first photovoltaic cells were measured by Becquerel, and others, around 1839. Copper-oxide- or silver-halide-coated metal electrodes were immersed in an electrolyte solution creating a so-called "wet photoelectrochemical" effect. To demonstrate this effect, one can place two copper sheets vertically in a glass and immerse them halfway in water containing copper sulfate or magnesium sulfate, which are electrolytes. After a few days, an oxide will form, and the illumination of one sheet in the solution will produce a small voltage that can be measured using a sensitive voltage meter connected to each sheet with alligator clips.

Modern solar cells, on the other hand, are not wet-photoelectrochemical solar cells. Such solar cells were first developed in 1954 at Bell Labs by D. M. Chapman and C. S. Fuller, using a solid-state semiconductor junction. These first silicon solar cells resembled those of today, but produced less than a watt of power. Today's PV solar panels are widely used to power satellites, and villages in third-world countries to produce power for buildings, and for utility-scale power generation. They are produced in quantities of several hundred megawatts (MW) per year.

Typical solar cells use a solid-state "P-N" junction that divides a region conducting a positive charge from a layer conducting negative charge carriers (i.e., electrons). This P-N junction is created by a multi-step process resembling that used by the semiconductor industry to manufacture integrated circuits and computer chips. A solar cell is actually a large-area diode since it conducts a charge in only one direction. Electrical charges created via the absorption of light in a semiconductor diffuse at different rates within the two types of layers in a solar cell, and are eventually collected and separated at the P-N junction. External contacts allow electrical currents to pass from the solar cell to the load. Individual PV modules produce direct-current (dc) electricity, and are available in 10-μW to 300-W sizes. Their actual power output depends on the intensity of sunlight, the operating temperature of the module, and other factors. Additional components such as electrical switches, diode-protection circuits, inverters, and batteries, connect the PV output with the electrical load. The resulting assembly of components is known as the PV system. There are numerous advantages to solar cells and systems, including their reliability, silence, long lifetime, low maintenance, flexibility, and low pollution during operation. Solar cells can also be incorporated as building materials in roofs and walls. Solar energy allows consumers to produce their own electricity and become more self-sufficient. Energy supply is then more easily distributed and less susceptible to breakdown. In general, solar energy can also help people have more control over their own energy supply and be less dependent on centralized sources of power. PV systems could provide "mini-utilities" to millions of homes. Although PV systems involve high-technology manufacturing, the assembly, installation, and maintenance of PV systems do not require a high level of skill or training. PV systems can provide local employment, especially in areas where it is most needed. This could play a part in helping a community achieve economic sustainability.

One of the perceived drawbacks with PV systems is that they are initially more expensive than traditional sources of electricity. However, government subsidies and incentives can help reduce costs. The effects of economies of scale are becoming increasingly more evident as costs for PV systems are dropping, but production costs need to continue to decrease. The cost of solar photovoltaics has dropped in price from over \$100 per peak watt in the 1970s to under \$6 per peak watt in 2002. Types of solar panels have diversified, and now include 100–200 W modules of crystalline silicon, amorphous silicon, cadmium telluride, and copper indium diselenide, among others. A new thin-film PV technology is based on photoelectrochemical solar cell technology and is called the dye-sensitized nanocrystalline solar cell (DSSC). It is based on organic dyes and is modeled after photosynthesis. The light-energy to electrical-energy conversion efficiency rate of commercial PV panels has typically ranged from 10% to 15%, with steadily increasing values over the last 20 years. With continued interest and investment, the trend in increasing efficiencies and decreasing costs is expected to continue. Growth in energy markets in the developing world has prompted energy giants such as British Petroleum, Kyocera, Siemens, and Shell to purchase PV manufacturers. At present, more than 100 MW solar panels are produced and shipped worldwide. The (2002) price per unit of solar-generated energy is approximately 30 cents per kWh—about three to five times the cost of conventional sources in most locations.

1.3 What is a Solar Cell?

There are two forms of solar conversion: (1) thermal conversion, where work can be extracted after being converted to thermal energy, and (2) quantum conversion, where the work output can be taken directly from the light absorber. In a thermal converter, the light is converted into heat at some temperature before work is extracted. An example of a solar-thermal converter is a solar hot-water heater. Another example might be a Stirling engine that is placed at the focus of a parabolic dish. In a quantum converter, a fixed number of photons yield a fixed number of energy "quanta" such as excited electrons. Examples of quantum converters include photographic plates, photosynthesis, vacuum photodiodes, and solar cells. The solar cell (see Fig. 1.1) is the focus of the discussion in this text, although all converters share similarities at the basic level.

In contrast to a thermal converter, in a quantum solar cell heat generation is actually to be avoided and is a sign that a portion of the energy is unavailable for extraction by an external circuit or load. A solar cell uses light-absorbing materials that generate what is called an electron-hole pair when the material is illuminated. The process is called "excitation of charge carriers by light." These carriers are separated in order to produce an external current through the load, resulting in electricity. From the standpoint of fundamental physics, the power output for all quantum solar converters, including the solar cell, is the product of the flow of the photo-induced product and the driving force for the flow. For a solar cell, the flow is the electrical current, and the driving force for the flow is directly related to the voltage. The voltage produced by the solar cell is dependant on the materials used, but it can also be understood by the fundamental optics of the device. These concepts are discussed more fully in Chapter 5. Since a voltage is produced from the action of photons, the term photovoltaic is used to describe the process in a PV solar cell. The solar conversion efficiency is determined by the output power divided by the incoming radiant solar power, while the total output current is limited by the number of absorbed photons, called the photon flux. We will examine the limitations of the current generation and collection in the solar cell from an optical standpoint in Chapter 2.

In the subsequent sections, each of these aspects is outlined for a typical solar cell. The focus is on the standard crystalline-silicon, or c-Si, solar cell. The concepts presented here are general and can be applied to any solar cell to gain useful insights into its design and improvement from an optical standpoint. The optical properties are described for the solar cell, and the basic equations that describe a solar cell's operation are presented. Finally, a more general model of a solar cell is presented that can be useful in understanding any quantum-conversion device from the optical aspect. Optical concentrators as well as the economic aspects of PV cells are discussed at the end of the tutorial.

Figure 1.1 shows a schematic diagram of a typical solar cell and the basic processes that occur during the photovoltaic effect. The solar cell device comprises two major regions, which are specially tailored to conduct negative and positive charges. An N-type material conducts electrons well, while the P-type material conducts a positive charge to a high degree. To understand how these materials differ, one must examine the materials themselves. The atoms in the solar cell are bonded to adjacent atoms in each material by their shared electrons. If an impurity, or dopant, such as a phosphorous atom is introduced into Si, it contributes an extra electron when it is incorporated into the structure, creating an N-type material. Likewise, if a boron atom is introduced, it is missing a bonding electron compared to Si and will create a P-type material. If a slab of P-type material is subsequently doped on one side with phosphorous, a so-called P-N junction is formed at the interface near the surface. Electrons from the donor atoms in the N region diffuse into the P region and combine with positive "holes" from the acceptor atoms, producing a layer of negatively charged impurity atoms. This process is shown in Fig. 1.2. The opposite effect also takes place when holes from the acceptor atoms in the P region cross to the N side, producing a layer of positively charged impurity atoms. The net result is an electric field that is positive on the N side and negative on the P side. It is this field in the "depletion area" or "barrier layer" at the junction that establishes the equilibrium, stopping the further movement of charges, and allows the device to act as a diode and as a solar cell.

Because of the presence of the junction, a photovoltaic cell acts as a rectifier, or diode, that allows easy passage of holes in one direction and electrons in the other. In order to understand why a photovoltaic cell acts as a rectifier, one can re-examine the energy-band diagram for the device. The diagrams shown in Figs. 1.2 and 1.3 represent the energy of the electron in the vertical direction, and the distance from the front to the back (right to left) of the device along the horizontal direction. A semiconductor is a material for which the allowed energy band for the highest energy bonding electrons is almost totally occupied. This is called the valance band (VB). It is separated from the set of energy levels for the excited electrons, the conduction band (CB), by the energy gap, also called the bandgap. This bandgap is 1.1 eV for Si and corresponds to a threshold wavelength of 1100 nm. If light of wavelengths shorter than 1100 nm enters a thick Si wafer, it is absorbed. This produces an electron in the CB, while leaving holes in the VB.

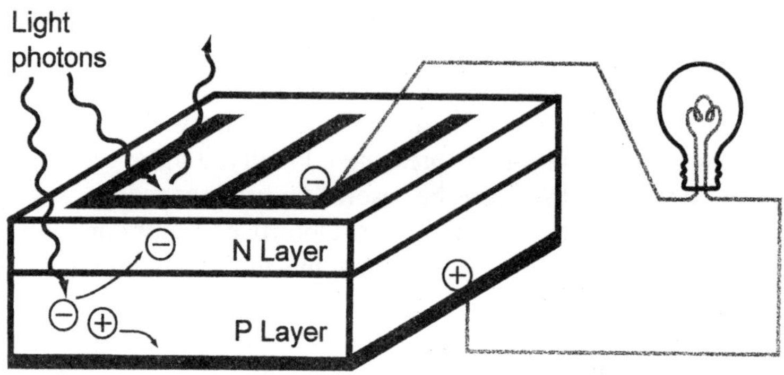

Fig. 1.1 A solar cell showing the processes of reflection of the incident light photons, light absorption by the semiconductor, free carrier generation via the absorption of light, and charge transport to the contacts. Electricity is produced in an external load (in this case a light bulb). A contact grid is shown on the N-type layer.

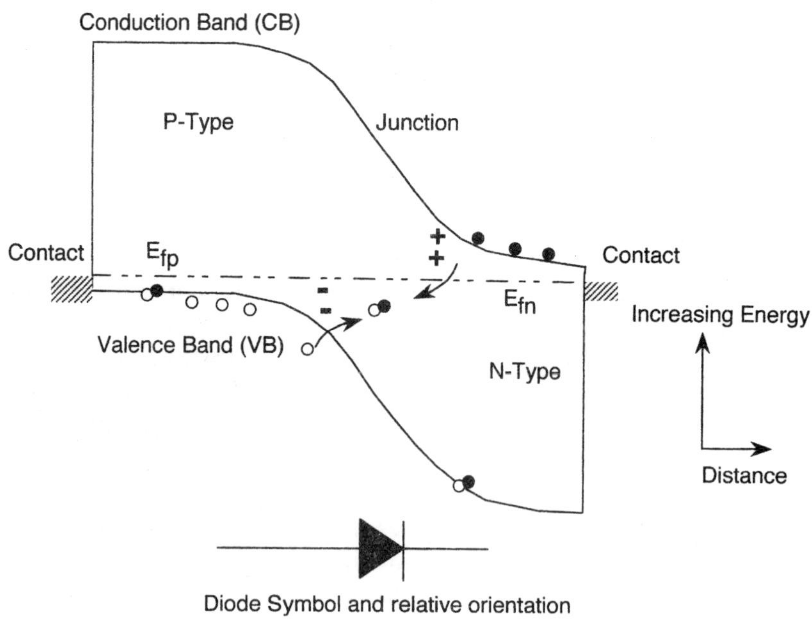

Fig. 1.2 This energy-band diagram for a P-N junction solar cell shows the formation of the built-in field at the junction. Holes are represented as unfilled circles, and electrons are shown as (dark) filled circles. Charge compensation is shown at the junction. The device is a rectifier, or diode, since external current passes easily only in one direction.

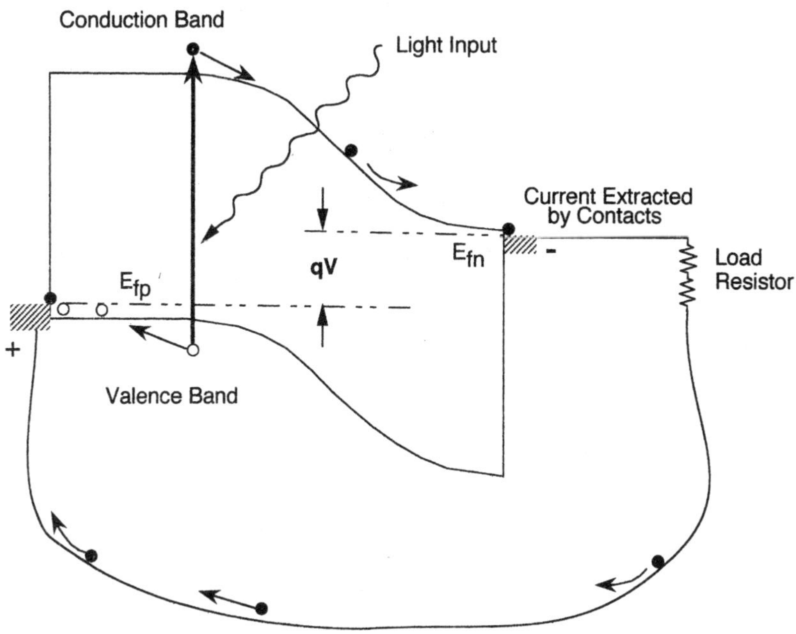

Fig. 1.3 An energy-band diagram for a P-N junction solar cell showing the generation and transport of charge carriers. Typically, either the N side or P side dominates the thickness of the device, and light enters through the side that has the contact grid.

When light enters the device and a photon is absorbed, an additional free electron is produced within these two layers. By "free" it is meant that the electron is no longer tightly bound to its host atom—it is mobile. What is left is a place in the light-absorbing material where an electron once resided. If charge neutrality was present before the light was turned on, the result is a positive charge at the site, called a hole, where the electron was ejected. Far from being fixed, this hole can also have "mobility." Electrons and holes are not only produced by light, they are also constantly produced (and destroyed) from thermal excitation. If a free electron within the device encounters the hole, it will fill it, and the hole will then reside on the site from where the electron came. It is not enough to have mobile electrons and holes. Ultimately, to produce work these charges must be separated and collected at the external contacts at the front and back of the device.

The diagram in Fig. 1.3 shows the energy-band diagram for a P-N junction solar cell, including the generation of charge carriers. Figure 1.4 shows the recombination of charge carriers. Electrons and holes created by the action of light are first elevated to higher energies. Instead of being collected by the

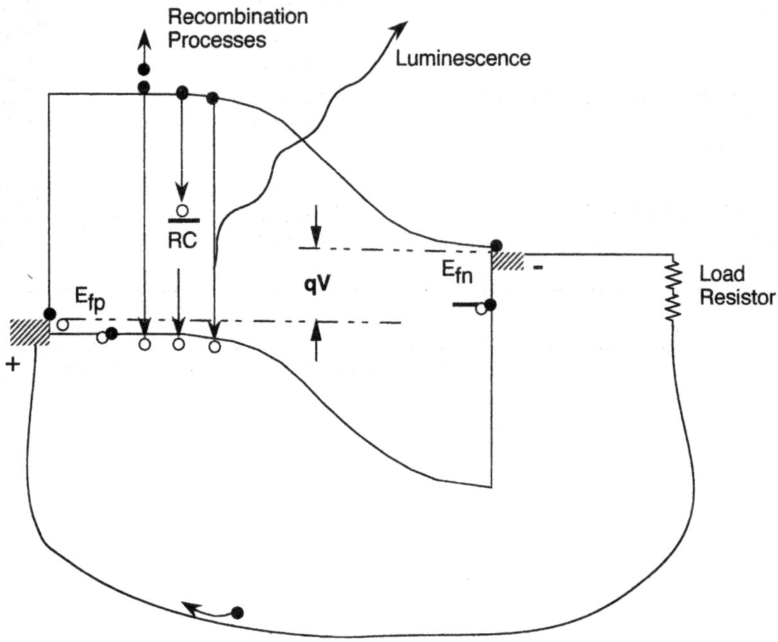

Fig. 1.4 Photoluminescent emission and nonradiative recombination compete with current extraction and power production in a solar cell.

junction and the external electrical contacts, these charge carriers can come back together, resulting in light (luminescence), or heat (nonradiative recombination). Photoluminescent emission and nonradiative recombination compete with current extraction and power production. Light is absorbed by the material to produce electrons in the CB, which can recombine in three distinct ways: radiatively, giving up the excitation energy in the form of an emitted luminescent photon; nonradiatively through traps, recombination centers (RC); or through excitation of CB electrons to higher levels producing a phonon or lattice vibration.

Alternatively, if the electron survives these recombination processes, it may be collected by the external circuit to produce a voltage, and to do work, as it recombines with the hole in the ground state or VB. This voltage, generated by the solar cell, is the difference between the chemical potentials of electrons and holes in the CB and VB, as shown by the dotted lines. These are the Fermi levels of holes in the P side, E_{fp}, and electrons in the N side, E_{fn}. This text shall use the convention that the energy is given in electron volts, eV. The voltage is then numerically equal to the change in chemical potential necessary to raise an electron through one volt (i.e., 1 eV). The photoluminescent emission from silicon solar cells can be predicted from optical data and a relationship called the Generalized Planck equation. This will be examined in Chapter 5.

In the dark, a condition called "equilibrium" exists where an equal amount of electrons created by thermal excitation are destroyed when they recombine with holes. In the dark, the current flowing to the right within the device is equal to that flowing to the left (see Fig. 1.2). Because of this, the voltage that develops at the external contacts is zero, and the Fermi levels are equal. When a P-N junction is used as a rectifier, the equilibrium is upset by the application of a voltage applied across the junction. There are other ways to upset that equilibrium; one of the most useful is to illuminate the device with light. As the light is absorbed, it ejects electrons from the bonds in the semiconductor, thus adding more electron-hole pairs to those produced by the thermal agitation. There are two viewpoints from which onc can scc how light produces the additional electron-hole pairs. Thinking of light as an electromagnetic wave, one can imagine that the oscillating electrical force of the wave pushes the bonding electrons rapidly back and forth because they bear an electrical charge. The resulting picture is somewhat like that of infrared light pushing chloride ions in a sodium-chloride crystal. Here, however, the light is oscillating a hundred times faster and shakes the electrons hard enough to tear them loose from their positions, leaving holes behind them. Alternatively, one can think of the light as a stream of particles, called photons, colliding with the electrons and knocking them out of position. However, it is best to fuse the two viewpoints—to picture light as a stream of wave bursts. In such a burst, a photon can exchange energy and momentum with an electronic wave burst when the wavelengths of the bursts are appropriate.

However, as the process is pictured, the light produces more electron-hole pairs than the semiconductor contains in its usual condition of thermal equilibrium. The additional free electrons on the N side of the junction and the additional holes on the P side are hardly noticeable, because their number is only a small fraction of the majority of carriers already present. Nevertheless, the holes added on the N side and the free electrons on the P side make a large fractional increase in the numbers of these "minority" carriers. By diffusing to the junction and sliding down the "hill" shown in Fig. 1.3, the added minority carriers increase the current flowing in one direction compared to that flowing at equilibrium. Therefore, a net electric current passes through the cell to the load resistor.

Thus, the operation of the solar cell depends on the junction capturing the free electrons drifting from its P side, but on that side they are greatly outnumbered by holes. The odds are good that an electron diffusing toward the edge of the hill will combine with a hole before it can escape through the junction. For this reason, the only useful electrons are those that are freed close enough to the junction to reach it before they are lost by combining with holes. In other words, the light that energizes the solar cell will be wasted if it is absorbed farther from the junction than the diffusion length, which is the average distance that an electron diffuses before a hole captures it. Solar cells are therefore designed as shown in Figs. 1.1–1.4. A solar cell is a large-area diode into which light can enter. Some wavelengths of light are only weakly absorbed by a given solar cell material. It is therefore necessary to understand the absorption properties of the solar cell material, and to tailor its thickness so that as much light as possible is absorbed. For Si cells, a thickness of about 200 μm is used in order to absorb the incident light efficiently. Typically, either the P-type or N-type region comprises most of this thickness

There are many types of materials used to create solar cells, the most common of which is crystalline Si. All solar cells share similarities at the fundamental level. One side of the device is conductive for electrons, and the opposite side of the device is conductive toward holes. Table 1.1 shows some of the various types of materials used in solar cells.

Table 1.1 Examples of some of the types of PV Cells. The hole conductor (hole cond.) in a DSSC is a material such as CuI, CuSCN, redox couple, or even a polymer, capable of conducting holes. This tutorial focuses on the c-Si device as an example.

PV Cell	N-type layer	P-type layer
Crystalline Silicon, c-Si	c-Si doped	c-Si doped
Gallium Arsenide, GaAs	GaAs	GaAs and AlGaAs
Amorphous Silicon, a-Si	a-Si doped	a-Si doped
Multicrystalline Si, Poly-Si	Poly-Si	Poly-Si
Cadmium Telluride, CdTe	CdS or ZnO	CdTe
Copper indium diselenide (-sulfide), $CuInSe_2$, ($CuInS_2$)	CdS or ZnO	$CuInSe_2$ or $CuInS_2$
Organic and polymer blend solar cells	Organic Molecule	Organic Molecule
Dye Sensitized Solar Cell, DSSC	TiO_2 + Dye	I^-/I_3^-, or hole cond.

One type of solar cell not listed above is the Schottky barrier device, in which only a P-type or N-type layer is used along with a metal or highly doped transparent conductive oxide (TCO). In this case, the band diagram looks like half of that shown in Figs. 1.2–1.4, either bending up or down at the interface, depending on which type of semiconductor is used. In polymer and organic solar cells, semiconductor particles and even fullerenes (C_{60}) are blended together to create a heterogeneous material that functions as a portion of the device. These materials can be used together, or a Schottky contact can be used. All of these various types of solar cells, and those not mentioned above, have to be incorporated into PV modules to be useful in a power-generating application, and all must have their current and power output characteristics measured as well.

1.4 Solar Cell Modules

Figure 1.5 shows a typical solar module. The solar cells are interconnected back to front (positive to negative) in electrical series to increase the voltage. In this way, typical PV panels of 12–16 V and 100 W can be produced. To make the

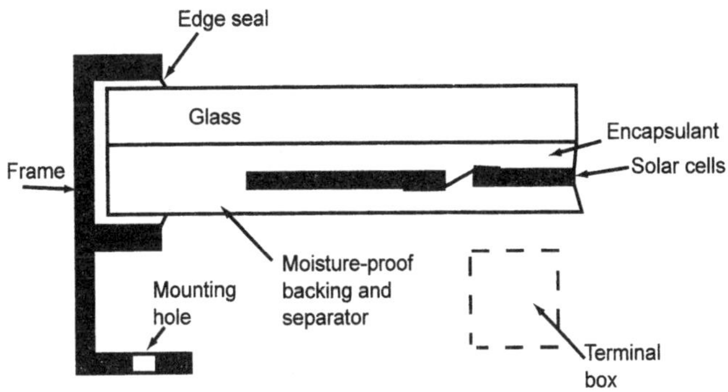

Fig.1.5 A PV module using individual solar cells. For example, the two solar cells shown can be individually (series) connected silicon cells.

panel, a tempered glass sheet is combined with a sheet of an encapsulant like silicone or ethylene-vinyl acetate (EVA) and laid into an assembly machine. Next, the series-connected string of solar cells is placed on top and a moisture-proof material, typically white Teflon or another suitable polymer, is applied to the back. The assembly is then placed in a vacuum press and heated to allow the encapsulant to flow, thus filling the voids and spaces. The final positive and negative side of the series-connected string of individual solar cells is allowed to exit the assembly so it can be tied to a terminal box for external electrical contact. The module is then sealed into a metal frame (typically Al) complete with mounting holes so that the module can be attached to rack or building assembly. Solar cells can also be interconnected in a series to create PV modules using techniques that resemble the manufacturing of integrated circuits. Figure 1.6 shows a typical "thin-film" solar cell that is fabricated on glass. The glass is coated with a thin film of a conductive transparent oxide such as SnO_2:F or indium tin oxide (ITO). The thin film is applied in long rectangular stripes that are isolated from one another by areas of bare glass. The solar cell is deposited next via techniques such as chemical vapor deposition (CVD), spray pyrolysis (SP), electrochemical deposition, or evaporation. A back metal contact is then deposited and the completed module can be mounted in a frame as in the previous case. Finished modules are typically measured for the I-V (current-voltage) curves to determine the power output.

A photovoltaic device can be modeled as an ideal diode in parallel with a light-induced current generator, I_{SC}. The short-circuit current, I_{SC}, is a function of the number of electron-hole pairs generated by the absorption of light that are collected. At this time, we can introduce the basic equation for the output characteristics of the solar cell. The current as a function of voltage, V, and the I-V characteristics of a solar cell are given by

$$I(V) = I_{SC} - \text{diode equation}\,. \tag{1.1}$$

The origins and derivation of this equation are explained in Chapters 3 and 5.

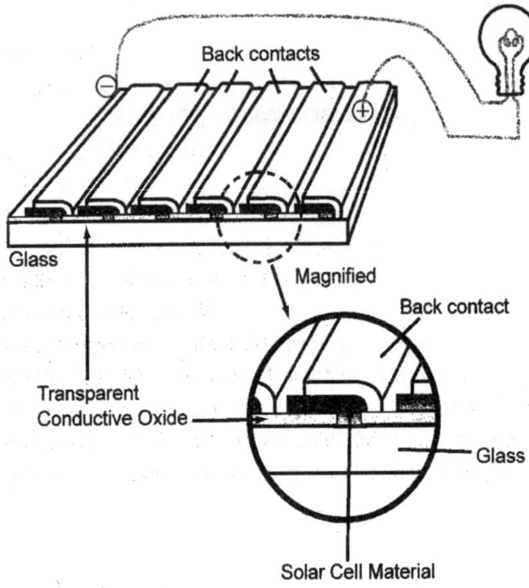

Fig. 1.6 A thin-film PV module configuration. The device is monolithically interconnected in series via the transparent conductive oxide (TCO). The solar cell materials (e.g., a-Si or CdTe) are sandwiched between the back metal contact and the TCO.

Figure 1.7 illustrates the results of the above equation plotted as an I-V curve and as a power-output curve. Note that when a solar cell is forward biased, it becomes a light-emitting diode (LED), and when it is reverse biased, it becomes a detector, or a photodiode, both of which are used extensively in optics and spectroscopy. For the photodiode, the photocurrent delivered to an external load is maximized, but the power output is minimal (or even negative). In the next chapter, the generation of this current via the absorption of sunlight is described. Then, the equations that describe the electrical output of the device will be outlined.

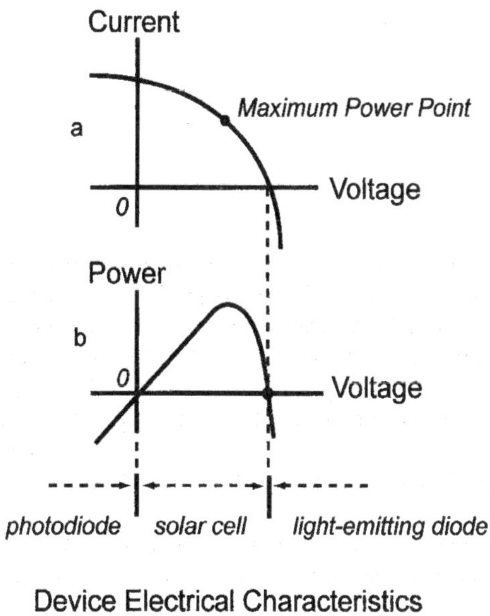

Fig. 1.7 Solar cell device electrical characteristics showing (a) I-V curve, and (b) Power output curve. These curves are typically expressed using the current density (e.g., mA/cm^2) vs. volts and the power density (e.g., watts/cm^2) vs. volts. A solar cell can be operated as a light emitting diode (LED) if it is forward biased, or it can be operated as a detector (photodiode) if it is reverse biased. In these cases, it consumes power instead of producing power (modified from DeVos [19]).

2

Absorbing Solar Energy

2.1 Air Mass and the Solar Spectrum

Now that we have introduced the solar cell, it is time to introduce the source of the energy—the sun. The sun has many properties that could be discussed at length. For example, the color temperature of the light, the nuclear (fusion) processes that occur within the sun, or the geometry of Earth and the sun that establishes the size of the solar disk as viewed from Earth. However, for the purpose of solar cell studies, two parameters are most important: the irradiance—that is, the amount of power incident on a surface per unit area—and the spectral characteristics of the light. The irradiance value outside Earth's atmosphere is called the solar constant, and is 1365 W/m^2. After being filtered through Earth's atmosphere, several portions of the solar spectrum diminish, and peak solar irradiance is lowered to approximately 1000 W/m^2. This is the typical irradiance on a surface, or plane, perpendicular to the sun's rays on a sunny day. If one were to track the sun for eight hours, the average daily solar irradiance would be approximately 1000 (8/24) = 333 W/m^2. On a fixed (nontracking) surface, the typical values in sunny locations range between 180–270 W/m^2. Solar data used for the purposes of PV-system sizing and economics are often expressed in units of insolation. The relationship between the average irradiance and insolation is given by the equation

$$\text{insolation}\,\frac{\text{kWh}}{\text{day}\cdot\text{m}^2} = \text{irradiance}\cdot\frac{24\,\text{h}}{\text{day}}\cdot\frac{10^{-3}\,\text{kW}}{\text{W}}. \tag{2.1}$$

For an irradiance of 250 W/m^2 the insolation would be 6 kWh/day/m^2.

The solar spectrum and irradiance is established by the air mass. Air mass (AM) refers to the amount of air a beam of sunlight must go through before reaching the solar converter. It is determined by the angle, θ, that the sun makes with a vertical line perpendicular to the horizontal plane (see Appendix, Fig. A.1). It is given by

$$\text{AM (number)} = \frac{1}{\cos\theta}\ . \tag{2.2}$$

The solar spectrum outside the atmosphere, AM0, is close to a 5743 K (Planck) blackbody-radiation spectrum and has an irradiance of 1365 W/m^2. The shape of this blackbody spectrum is shown in Fig. 2.1. Air mass 1.0 refers to the thickness of the atmosphere sunlight passes through if the beam is directly overhead. An AM1 atmosphere reduces the direct flux by a factor 0.7. On a clear day, and when the sun is directly overhead, nearly 70% of the solar radiation incident to Earth's atmosphere reaches its surface undisturbed. About another 7% reaches the ground in an approximately isotropic manner after scattering from atmospheric molecules and particles. The rest is absorbed or scattered back into space. Both the direct and scattered fluxes vary with time and location because the amounts of dust and water vapor in the atmosphere are not constant even on clear days. For purposes of standard solar cell measurements, an average solar spectrum at AM1.5 is used ($\theta = 48.19$ deg). It should be noted that the total irradiance used for AM1.5 was 844 W/m^2 in earlier work, but is often normalized to 1000 W/m^2 in more recent work (ASTM E 892, IEC 60904-3). It is therefore best to specify the AM and the irradiance when reporting measurement conditions. Figure 2.1 shows the spectral irradiance for the sun when viewed as a blackbody, but is scaled (diluted) such that the total power is approximately 1000 W/m^2.

An attempt to replicate the AM1.5 spectrum is made in standardized solar simulators. Figure 2.2(a) shows the solar spectrum at AM1.5 (see Appendix, Table A.1). The integral over the wavelength yields the total irradiance, 1000 W/m^2. The many notches in the spectrum are attributed to the absorption bands of various atmospheric gases such as H_2O, CO_2, O_3, and O_2. Absorption by ozone is essentially complete below a wavelength of 0.3 μm. The relatively large attenuation below 0.8 μm is due to scattering of molecules and particulates. These scattering processes become weaker at longer wavelengths, as has been shown by both theory and observation. This also explains the spectrum of the diffuse radiation, which is richer than the direct radiation in the blue portion of the spectrum. When analyzing the performance of solar cell systems, the cell output is usually assumed to be proportional to the solar radiation intensity with little regard to the variations in the spectral distributions. This practice is satisfactory for engineering and design purposes, but may be problematic for reporting accurate solar conversion efficiencies. The amount of cloud cover is a dominant factor in determining the transmission and scattering of solar radiation in practical PV applications.

The solar spectrum discussed above can be used to determine the number of photons that can produce electrons in the solar cell. The wavelength scale on the solar spectrum can be converted to photon energy, e, from the relationship

$$\text{Photon Energy} \equiv e = \frac{hc}{\lambda} \cong \frac{1.239}{\lambda(\mu m)} \text{ [in eV]}. \tag{2.3}$$

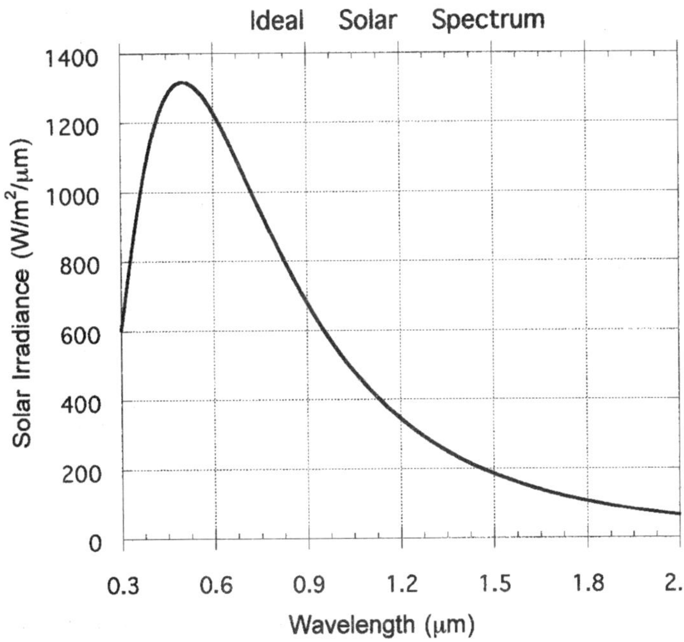

Fig. 2.1 Solar spectrum obtained from the Planck blackbody equation and T_s= 6000 K. The plot is normalized so that the total power (area under the curve) is 1000 W/m^2.

Thus, a photon at a wavelength of 550 nm has an energy of approximately 2.2 eV. Knowing the energy per photon at each wavelength, the *y* axis in the solar spectrum shown in Fig. 2.2(a) can be converted to a number of photons per second per unit area and photon energy. Such a plot, corresponding to Fig. 2.2(a), is shown in Fig. 2.2(b). To convert the solar irradiance, P, one uses the relationship

$$\frac{d\Gamma}{de} \equiv \Gamma_e = \frac{d\Gamma}{d\lambda} \cdot \frac{d\lambda}{de} = \Gamma_\lambda \frac{\lambda}{e} = \frac{P_\lambda}{e} \cdot \frac{\lambda}{e}, \tag{2.4}$$

where e is the photon energy and Γ is the photon flux density. A plot of photon-flux density such as the one shown in Fig. 2.2(b) is useful in establishing the limits on the photocurrent from a solar cell. The expected maximum current can be calculated if the number of absorbed photons per unit area is multiplied by the charge per electron, q. For example, if a solar cell could absorb all photons with an energy of 1.6 eV and higher, and each photon created an electron collected by the external contacts, then a current of approximately 20 mA/cm^2 would result in the external circuit.

As another example, consider a Si wafer of a 200-μm thickness and a 1-cm^2 area illuminated by AM1.5 sunlight. As a first approximation, we can consider all the charge carriers to be uniformly distributed within the volume of the solar

(a)

(b)

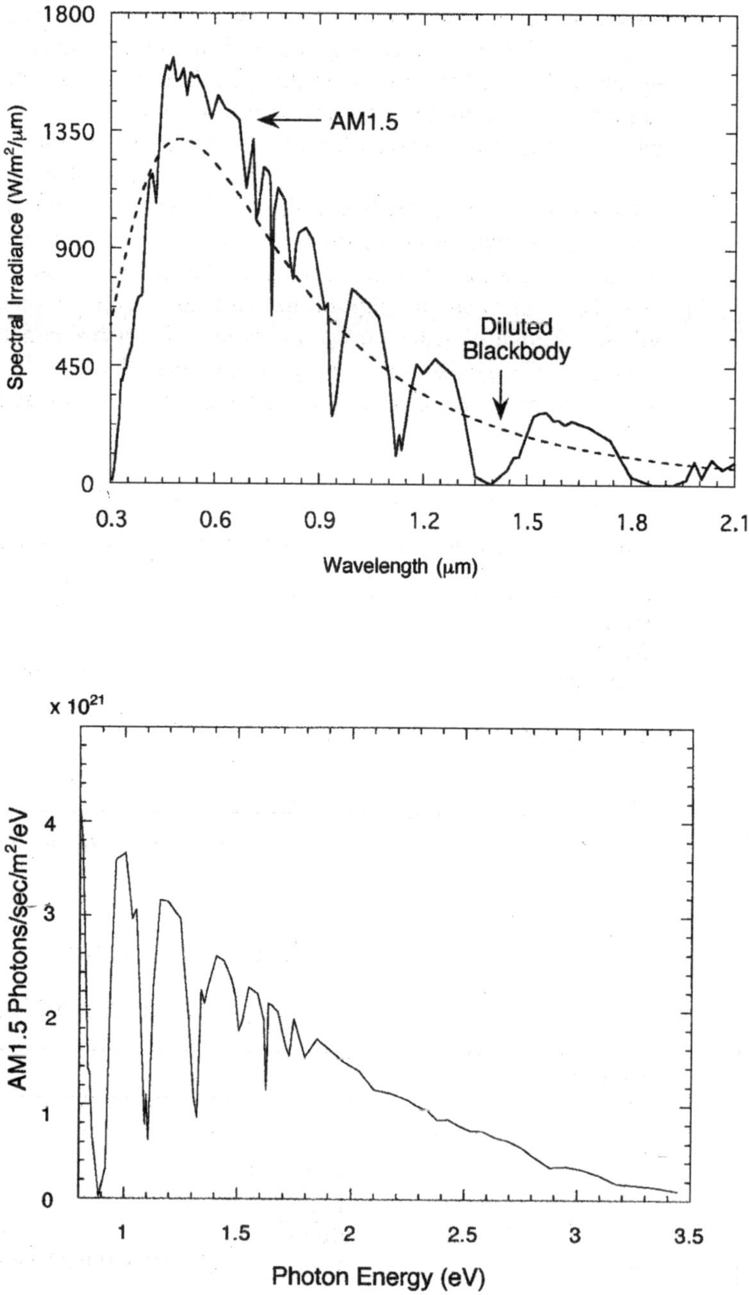

Fig. 2.2 The solar spectrum at AM1.5, 1000 W/m^2 conditions for (a) Irradiance normal to the beam, and (b) corresponding photon flux (number of photons). For the AM1.5 data, see the Appendix, Table A.1.

cell. If the J_{SC} value for a solar cell made with Si is approximately 30 mA/cm^2, the total incident photon flux density [integral of Fig. 2.1(b)] is greater than $J_{SC}/q = 1.9 \times 10^{17}$ photons/sec/cm^2, and the excited electron concentration is $J_{SC}/(q \times 200 \times 10^{-4}$ cm$) = 9 \times 10^{18}$ electrons/sec/cm^3. Of course, the steady-state electron concentration (in electrons/cm^3) present in the solar cell at AM1.5 would be much lower than this value. This is because the charge carriers are swept out of the device so that they can be collected by the contacts to flow through the load. An analogy would be to ask how many cars are present on a busy section of road at a given moment versus how many cars have passed a point on the road per hour. In Chapter 3, we will see that the electron concentration in a solar cell is obtained from a balance between the number of charge carriers produced by photon absorption, and the subsequent charge-carrier recombination and diffusion in the light-absorbing material.

2.2 Optical Properties of Solar Cell Materials

2.2.1 Absorptivity

Just how many photons can be absorbed by a solar cell is determined by the optical properties of the device, which in turn, is a property of the material used to absorb the light and the geometry in which it is used. One of the most fundamental questions in solar cell design and analysis is whether much of the solar spectrum can be absorbed. There are several useful optical parameters to be considered when characterizing a solar cell or solar cell material. Some are fundamental constants of the material, others are "lumped" parameters that only characterize the particular device or solar geometry in question. Listed in the order of most to least fundamental (basic), the constants are the complex index of refraction, the extinction coefficient, the absorption coefficient, and the absorptivity. In this section, we shall examine each of these to demonstrate their interrelationship and connection to solar cell design.

When determining a solar cell's light absorption, it is the optical parameter called the absorptivity that is most useful when assessing potential absorber materials for solar cells, or when optimizing a given absorber material for a solar cell. The quantum absorptivity is the fraction of the incoming light at a given photon energy, e, that is absorbed by the material to produce an excited state such as an electron-hole pair. It is measured, and calculated, as a function of the photon energy, yielding a(e), or, alternatively, it can be expressed as a function of the photon wavelength, yielding $a(\lambda)$. The absorptivity can be multiplied by the incoming photon flux to determine how many electron-hole pairs can be produced. Multiplying this result by the elemental charge, q, and integrating over the solar spectrum then yields the upper limit for how much current can be extracted from a device made with the solar-absorber material. The absorptivity can be measured directly, or it can be calculated using the basic optical properties that are constant for a material. Not all absorption in a solar cell material creates

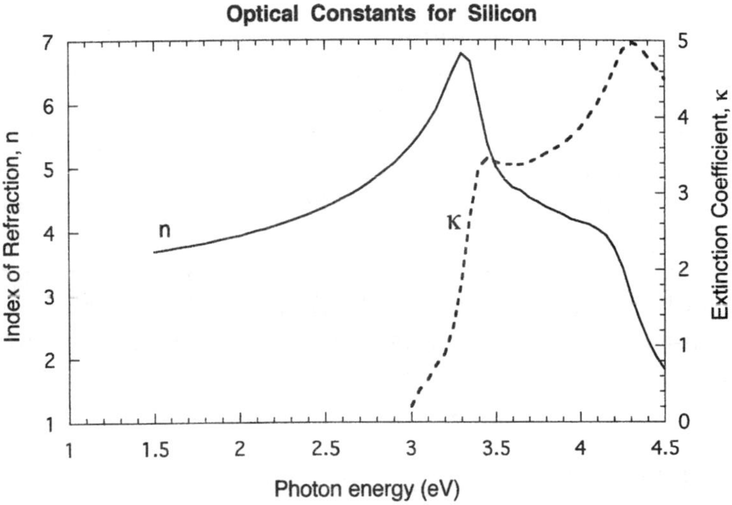

Fig. 2.3 Optical constants n and κ for silicon as real and imaginary components of the index of refraction. At photon energies smaller than 3 eV, the extinction coefficient of Si is below 0.006, and the index of refraction, n, is approximately 3.5.

electron-hole pairs. For example, absorption by free carriers (e.g., electrons) can themselves contribute to the total absorptivity, although they ultimately produce only heat (e.g., lattice vibrations) and no additional photo-excited species (e.g., electrons). As a first approximation for solar cell measurements, the total absorptivity can often be used in place of the quantum absorptivity. An umderstanding of the basic optical properties of a solar cell material is used to determine the absorptivity and the optimum thickness for a solar cell device. This can be determined from transmission and reflection measurements. We shall now examine the fundamental optical properties for a solar cell material and connect these with the absorptivity.

2.2.2 Absorption coefficient

The optical properties of a material depend on the complex index of refraction, n_c, given by the relationship

$$n_c = n - i\kappa , \tag{2.5}$$

where the imaginary part of n_c is the extinction coefficient, κ. The real part of n_c is the index of refraction, n, familiar in optical design of lenses and geometrical optics. Note that the symbol "n" is used in most texts for both electron concentration, to be discussed later, and index of refraction. Figure 2.3 shows the n and κ values for silicon. Typically, n_c, n, and κ are determined by using a technique called ellipsometry. This is an optical-based technique for the in-situ nondestructive characterization of interfaces using the change in a light probe's

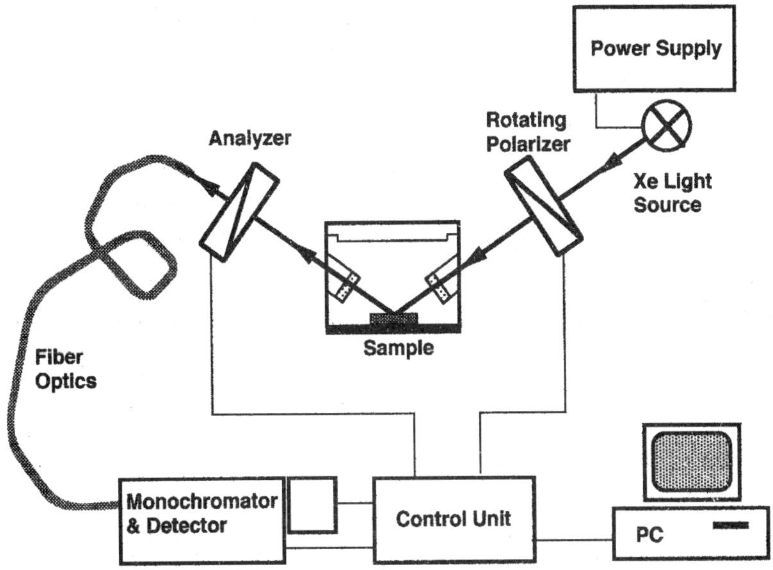

Fig. 2.4 A typical ellipsometer setup used to determine n and κ.

state of polarization. A diagram of the setup is shown in Fig. 2.4. It relies on the fact that linearly polarized light is elliptically polarized when reflected from a dielectric material, such as a semiconductor, used in solar cells. In addition to establishing the basic optical properties of a solar cell material, this technique can also be used to estimate the depth profile of impurities. Many commercial ellipsometry systems exist for these purposes.

The extinction coefficient is related to the absorption coefficient by the relationship

$$\alpha(\lambda) = \frac{4\pi\kappa(\lambda)}{\lambda}. \qquad (2.6)$$

Figure 2.5 shows the absorption coefficient of Si as a function of wavelength $\alpha(\lambda)$. The absorption coefficient, n, and κ are all a function of the wavelength of the light, or can alternatively be expressed versus the photon energy, e. For example, for silicon at a wavelength of approximately 800 nm (e = 1.55 eV), $\alpha = 10^3$ cm^{-1}, and so $\kappa = 0.006$. It should be pointed out that the absorption coefficient for a material, although a function of wavelength, is not affected by the thickness of the wafer. However, the optical absorption is a strong function of the thickness and the geometry of the solar cell.

The absorption coefficient describes the decrease in light intensity as a beam of light propagates through a material (e.g., a solar cell). If the number of photons per unit time per unit area is Γ, then the change in this "photon-flux density" as a function of position is given by

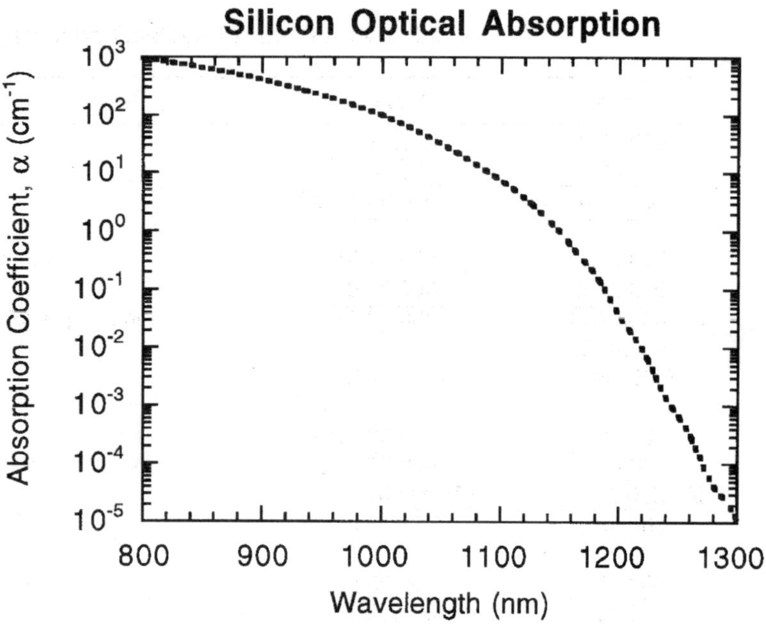

Fig. 2.5 Silicon optical absorption coefficient, α.

$$\frac{d\Gamma}{\Gamma} = -\alpha \, dx, \qquad [2.7(a)]$$

where x is the position in the absorbing material along the beam. Here, reflection is neglected, and only the light that has already entered the light-absorbing material is considered. By integrating the above equation, one obtains the number of unabsorbed photons. This is given by

$$\Gamma = \Gamma_0 \exp(-\alpha t), \qquad [2.7(b)]$$

where t is the thickness of the material, and Γ_0 is the number of photons that initially entered the material. The transmission, T, is therefore equal to Γ/Γ_0. This equation is one form of the Beer-Lambert law. Many are familiar with this equation as it applies to the absorption of solutions, which, for example, could be a colored dye dissolved in alcohol. In this case, a molar-extinction coefficient, ε, and a concentration, C, are used in the decadic form of the above equation,

$$T = 10^{-\varepsilon C t} = 10^{-OD}. \qquad [2.7(c)]$$

If the path length is measured in centimeters, the units of ε are typically $M^{-1}\,cm^{-1}$, and those of C are M^{-1}. The term M, familiar in chemistry, is molar, or moles per liter. The exponent is given the term absorbance, optical density, or OD for short. As can be seen, the absorption coefficient, α, is equivalent to 2.303 ε C. This is useful, for example, in studies of organic solar cells, where the primary light absorber is a dye or organic molecule. It is customary to measure the molecule in a solution of known concentration, C, and then compare its absorption coefficient value and absorbance versus wavelength after it has been deposited as a solid. Due to varied interactions between the molecules in a solution versus a solid, the absorption coefficient in a solution is often different than that of a solid film of the same material. The above equations are useful for communicating the results of these cases.

2.2.3 Solar cell bandgap

Another important optical parameter for solar cells is the "optical bandgap" of the solar cell-absorber material. This is the minimum energy necessary to elevate an electron to the excited state, or upper energy level, so that it can be conducted through the solar cell to the load. For Si, the bandgap energy is 1.1 eV, so the corresponding wavelength is 1100 nm. Only when the thickness of the Si wafer approaches several hundred microns do the apparent optical transition and the bandgap converge. This will be illustrated shortly. Determination of the optical bandgap can establish the upper bounds for the solar conversion efficiency of a PV solar cell. Too large a bandgap, and the material will only absorb short wavelengths of light (high photon energy, e), thus the device will be limited to a small photocurrent under AM1.5 illumination. Too small a bandgap, and the solar cell can produce a large photocurrent, but a small voltage and low efficiencies result. The optimum optical bandgap for the AM1.5 solar spectrum is approximately 1.35 eV. The basis for this will be discussed in Chapter 5.

The absorption coefficient can be used to establish the bandgap, E_g, of the absorber material used in the solar cell. In "direct-bandgap" solar cell materials, such as GaAs and CdTe, this optical transition does not require assistance from the energy of lattice vibrations called phonons. These phonons are the vibrations of the host atoms in the light-absorbing material (e.g., GaAs, or Si). The relationship between the absorption coefficient and the bandgap for direct-bandgap materials is

$$\alpha(e) = C_d \left(e - E_g\right)^{1/2}, \qquad [2.8(a)]$$

where C_d is a constant that is approximately 2×10^4 if α is given in typical units of cm^{-1}. This equation can be derived from simple quantum mechanical considerations and conservation of electron energy and momentum. The usefulness of this equation is that a plot of α^2 versus photon energy yields an *x*-intercept that is the bandgap of the material. For GaAs, for example, this yields a value of 1.42 eV. Because of their strong absorption, even thin layers (e.g., 10 μm) of direct-bandgap materials can absorb more than 90% of light with energies larger than their bandgap energies. This is an advantage in solar cell devices, and leads to the concept of "thin-film" solar cells.

In contrast to direct semiconductors, phonon assistance is required for so-called "indirect-bandgap" materials such as Si or Ge. The result of this is that the spectral absorption of the material gradually tapers off near the bandgap wavelength instead of being a sharp transition (see Fig. 2.5). In this case, the absorption is of the form

$$\alpha(\mathrm{e}) = \frac{C_i\left(\mathrm{e} - E_g + E_p\right)^2}{\exp(E_p/kT) - 1} + \frac{C_i\left(\mathrm{e} - E_g - E_p\right)^2}{1 - \exp(-E_p/kT)}. \quad [2.8(b)]$$

The data shown in Fig. 2.5 was used to plot $\alpha^{1/2}$ versus photon energy shown in Fig. 2.6. There are two extrapolated intercepts, and the midpoint between the two yields the indirect bandgap, E_g. The distance from E_g to the intercepts yields the phonon energy, E_p, for the material. This is typically in the range 0.01–0.08 eV. A common error for the characterization of new materials is to perform such an analysis and conclude that a material has an indirect bandgap with a phonon assistance of more than 0.5 eV—an unlikely number for a lattice vibration. Finally, it should be noted that both the "constants" C_i and C_d have in their denominators a factor of the photon energy, e, and so $(\alpha e)^{1/2}$ and $(\alpha e)^2$ are often plotted, although the corresponding difference in the E_g values determined with the inclusion of the photon energy is negligible.

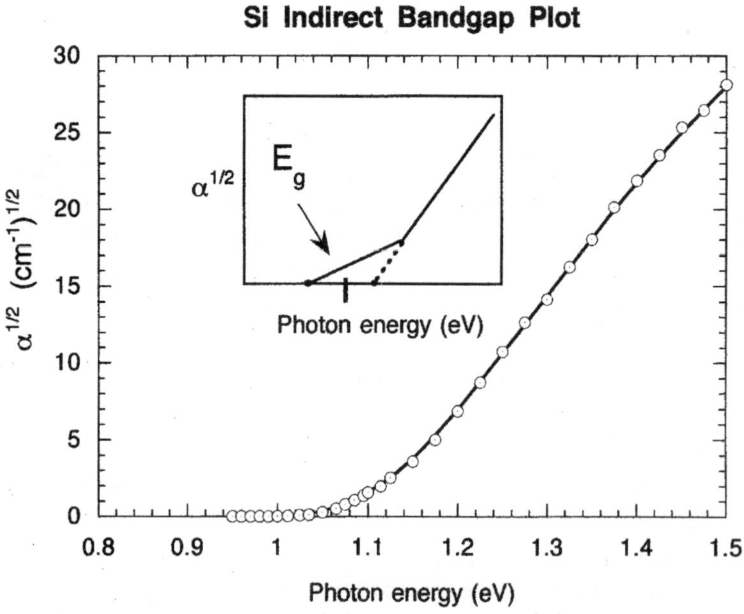

Fig. 2.6 An indirect bandgap plot of the data for Si in Fig. 2.5. This indicates a bandgap of 1.1 eV from the midpoint between the two extrapolated x-axis intercepts.

The general experimental setup for these measurements is shown in Fig. 2.7(a), which uses a double-beam spectrophotometer and a broadband-integrating sphere to capture any scattered or diffuse light. Light is absorbed, reflected, or transmitted; therefore, the absorptivity can be determined from the direct measurement of reflectivity, Ref, and the transmission coefficient, T, from the relationship

$$a\,(e) + \mathrm{Ref}\,(e) + \mathrm{T}\,(e) = 1. \tag{2.9}$$

Figure 2.7(b) shows a close-up of a typical integrating-sphere setup to determine transmission (T) + reflection (R) versus wavelength curves for a wafer. This setup has been used to determine the T + R values of polished Si wafers of various thickness. The results from this measurement are illustrated in Fig. 2.8, and demonstrate that the apparent absorption edge may appear to shift several hundred nanometers in wavelength for the same absorber material (e.g., Si). Far from being an academic issue, the 12-µm Si wafer is expected to produce less photocurrent (i.e., amperage or I_{SC}) than the 2.75-mm Si wafer when they are

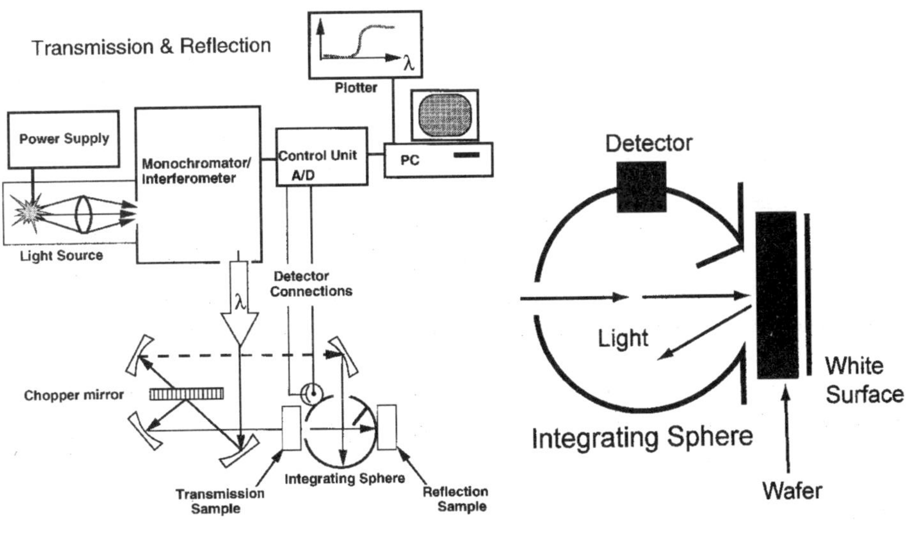

(a) (b)

Fig. 2.7 (a) Typical experimental setup for the measurement of solar cell (or solar cell materials) optical properties. A/D is an analog-to-digital converter. A grating monochromator, or an interferometer (for FTIR measurements) can be used to produce monochromatic light of a given wavelength that can either be transmitted through or reflected from a sample. Any scattered light can be collected via an integrating sphere. (b) Close-up showing the integrating sphere and detector in the geometry used to measure T + R for the geometry relevant to solar cells.

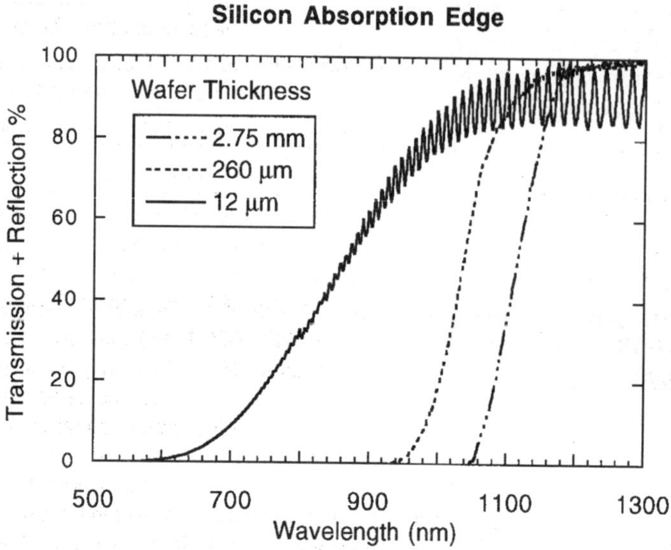

Fig. 2.8 Sum of the hemispherical transmission and reflection (T + R) vs. wavelength measured for three silicon wafers of different thickness.

used to fabricate solar cells. As illustrated in Fig. 2.8, red light can be observed upon viewing a white light source through a thin (12-μm) Si wafer. This would mean the light may be lost to the system and not used to produce electron-hole pairs. Being able to predict and understand the optical properties of a solar cell is therefore critical in PV design.

2.2.4 Antireflection coatings

One of the critical design aspects of solar cells is the optimization of the absorptivity by methods that include antireflection coatings, as well as techniques that force the light to travel long pathways within the fabricated solar cell device. The latter involves texturing the front or back surface of the cell, a topic to be discussed shortly. Figures 2.9(a) and 2.9(b) show the transmission, reflection, and calculated absorptivity from a standard and Fourier transform infrared (FTIR) measurement performed on a 260-μm-thick silicon wafer coated with a thin SiO_2 layer. The setup for an FTIR-based transmission or reflection measurement is identical to that in Fig. 2.7(a), except that an interferometer is used to produce monochromatic light instead of a grating-based monochromator. Figure 2.9(b) shows the wavelength range extended into the IR. This shows the presence of a defect, dopant, or impurity absorption at approximately 9 μm, and the free-carrier absorption (i.e., $\lambda = 10 - 14$ μm) mentioned earlier. Again, it should be pointed out that such absorption, at wavelengths beyond the bandgap (i.e., $\lambda > 1100$ nm for Si), is unlikely to produce electron-hole pairs that lead to a photocurrent in an external circuit. This absorption is, however, useful in characterizing the extent of doping, or the quantity of impurities.

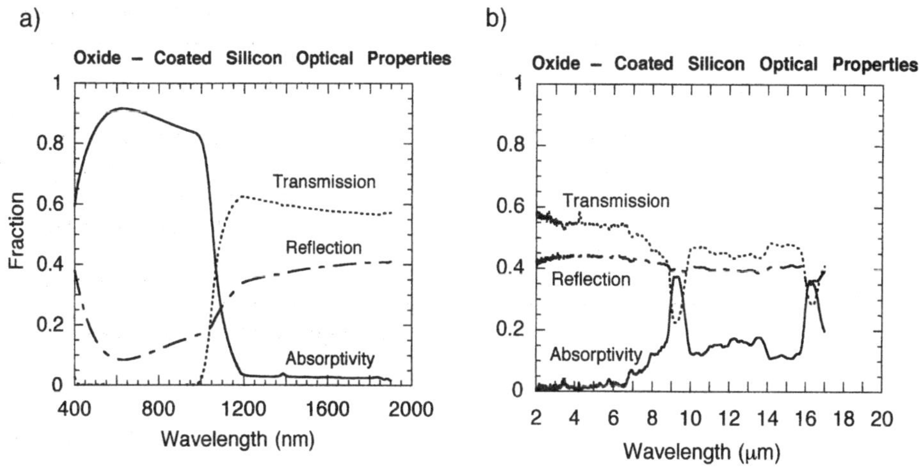

Fig. 2.9 Results from an FTIR based measurements of a silicon wafer in the visible and near IR (a) and far IR (b). Although the absorption edge for silicon is near the expected wavelength (1.1 μm), additional absorption in the IR indicates that band-to-band excitation is not the only transition possible. Since the wafer is coated with SiO, an antireflection effect is seen in the reflection spectra near 600 nm.

A bare silicon wafer can reflect as much as 30% of the incoming light from the sun in its useful absorption range (350–1100 nm). If unabsorbed, this light cannot be used to generate electricity. Figure 2.9(a) shows, in contrast, that if the Si wafer is covered with a thin layer of the "native" oxide (SiO and SiO_2), the reflectivity can be reduced to approximately 10% for some wavelengths. With more sophisticated coatings consisting of such materials as TiO_2, Al_2O_3, or Ta_2O_5, this reflection loss can be further reduced to less than 3%. A thin-film coating can be grown by heating the wafer (to produce SiO and SiO_2), or by deposition via sputtering, evaporation, or sol gel techniques, and it can serve as an antireflection coating (ARC) for a solar cell fabricated from a wafer. This ARC can be understood from basic optical principles.

For a simple slab of material like that shown on the left side of Fig. 2.10, the reflectivity at normal incidence (orthogonal to the wafer) is, to first approximation, given by the index of refraction of the slab, n, and the index of refraction of the surrounding medium, n_0, from

$$\mathrm{Ref}_1 = \frac{(n-n_0)^2}{(n+n_0)^2} \quad . \tag{2.10}$$

The full form of this equation is given by

$$\mathrm{Ref}_1 = \frac{(n-n_0)^2+\kappa^2}{(n+n_0)^2+\kappa^2} \quad . \tag{2.11}$$

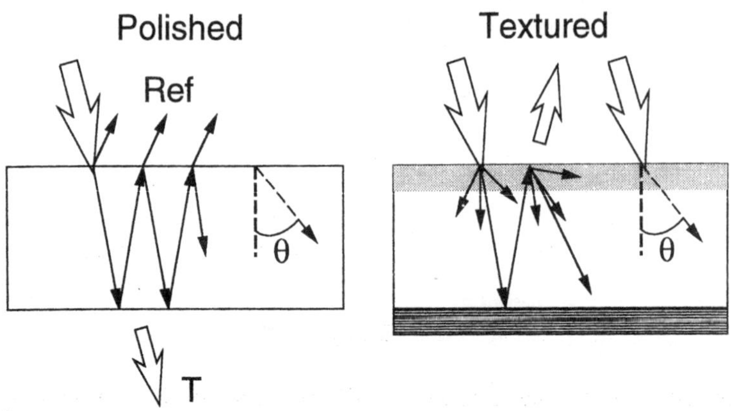

Fig. 2.10 Geometry for calculation of the absorptivity from a polished (left side) and textured (right side) light absorber. In each case, a summation of multiple reflections must be made in order to calculate the absorptivity. The angle θ is measured from the surface normal.

These equations can be used to predict the reflectivity, otherwise known as the reflection coefficient, of a solar cell absorber material. The index of n value for Si, for example, is approximately 4 at a wavelength of 600 nm (see Fig. 2.3). For a wafer in air ($n_0 = 1$), this yields a value for the reflectivity value between 30 and 40%. In other words, only about 70% of the light enters a bare Si wafer. If nothing were done about this loss, a solar cell would be impractical and inefficient.

Instead, if a thin ARC is deposited on the Si wafer, for example, reflection can be minimized at certain wavelengths. This is understood by recalling that a beam of light undergoes a phase change of π (180 deg) when passing from a medium with lower index of refraction to one with a higher index of refraction. No phase change occurs from a medium of higher n value to a lower n value medium. A portion of the incident beam is not reflected; it is transmitted at the back surface at the ARC-Si interface and reflected from there (see the left side of Fig. 2.10). If the thickness of the ARC, t_{ARC}, allows an optical path $2t_{ARC} = 1/2\ \lambda/n$, then destructive interference occurs between the two beams, and the light enters the Si wafer across the ARC with little reflection loss. This condition is expressed as

$$n_{ARC}\, t_{ARC} = \lambda / 4, \tag{2.12}$$

where n_{ARC} is the index of refraction of the ARC (e.g., n SiO_2 or SiO = 1.4–1.5, n TiO_2 = 2.3, and n Si_3N_4 = 1.8–1.9). For example, a reflection minimum at a wavelength of 600 nm requires a 100-nm thickness of SiO_2, while only 65–70 nm is required for an ARC if TiO_2 is used. This so-called "quarter-wave" coating allows for a minimum in the reflection coefficient described by

$$\text{Ref}_{\text{min}} = \frac{\left(n_{ARC}^2 - n_0 n\right)^2}{\left(n_{ARC}^2 + n_0 n\right)^2} . \tag{2.13}$$

The lowest (i.e., optimal) Ref_{min} value occurs if the condition

$$n_{ARC}^2 = \sqrt{n_0 n} \tag{2.14}$$

is satisfied. Since the index of refraction for Si is a function of wavelength, an optimum ARC thickness can only be selected for a single wavelength. Antireflection coatings are typically designed for the peak in the solar spectrum, a wavelength of approximately 600 nm. For example, a 70-nm-thick coating of TiO_2 on a Si wafer results in a single minimum in the reflection coefficient, Ref_{min} = 1–2%, at 600 nm, with an overall reflection coefficient of approximately 10% in the range of 350–1100 nm. The coated Si wafer reflects more UV and blue light than red light [see Fig. 2.9(a)]. This is why Si-based solar cells often take on a blue appearance. A two-layer ARC consisting of 70 nm of TiO_2 and 110 nm of MgF_2 results in minima in reflection coefficient at both 475 nm and 875 nm, and results in overall reflection losses of approximately 3% under solar illumination. The ARC technique can also be used to produce wafers with various colors or optical properties if this is desired for consumer applications and if this aspect is considered more important than over maximizing the overall conversion efficiency.

2.2.5 Thickness determination

Interference effects can also be used to learn about the solar cell material itself. For the 12-μm Si wafer shown in Fig. 2.8, interference fringes are observed as expected for thin films. This is often observed for so-called thin-film solar cells, examples of which are CdTe, $CuInSe_2$, and amorphous Si (a:Si). Typically, interference fringes are most easily observed at wavelengths longer than the bandgap wavelength. A maximum in reflection is expected at wavelengths that satisfy the condition

$$2\,t = \left(m + \frac{1}{2}\right)\frac{\lambda}{n}, \tag{2.15(a)}$$

where m is an integer (0, 1, 2, 3…..). By measuring the wavelengths (λ_m and λ_{m+1}) at which two adjacent maxima occur in a reflection spectra, such as shown in Fig. 2.8, it is possible to determine the thickness, t, of a material given its index of refraction, n, or visa versa. The relationship

$$t = \frac{1}{2n}\left(\frac{\lambda_m \lambda_{m+1}}{\lambda_m - \lambda_{m+1}}\right) \tag{2.15(b)}$$

can be used to determine a thickness of a thin film deposited on a supporting substrate. For example, for the plot shown in Fig. 2.8, the reflection spectra dominate the pattern (structure) of the interference, since the transmitted beam is scattered. Using λ_m and λ_{m+1} values of 1157.6 nm and 1141.5 nm, respectively, from the plot, the thickness of the Si is determined to be approximately 12 μm (for Si, n = 3.6). If the transmission spectra instead of the reflection of a thin film are measured, then the factor of 2 in the denominator of the above equation is omitted. This allows the thickness of thin-film solar cell materials such as amorphous Si, CdTe or $CuInSe_2$ to be determined on substrates such as glass. The thickness of an ARC or the thickness of an absorber used in a thin-film solar cell can thus be determined from a measurement of the interference fringes.

2.2.6 Predicting absorptivity

Turning back to the light absorption in the solar cell, the absorptivity can be predicted from the absorption coefficient, α, and the thickness, t, given the geometry of the light absorber. As an example, the absorption-coefficient data for silicon can be used. Most relevant for solar cell applications is the case of a simple parallel-plane solar cell geometry that is a slab of absorber material. This is the case of a semiconductor wafer. For a planar geometry and a polished wafer, combining the Beer-Lambert equation together with a treatment of multiple reflections, one can predict the absorptivity (see the left side of Fig. 2.10). If the reflection coefficient for a single reflection event is Ref_1, and we use the absorption coefficient at each photon energy, $\alpha = \alpha(e)$, then the absorptivity for light of near-normal incidence on a parallel slab of material is given by

$$a(e) = [1-\exp(-\alpha t)]\frac{1-\mathrm{Ref}_1}{1-\mathrm{Ref}_1\exp(-\alpha t)} . \quad (2.16)$$

In contrast to the simple planar case, the radiation can undergo multiple diffuse scattering as would be the case if the light is incident on a textured or roughened interface (see the right side of Fig. 2.10). Solar cells are often textured so that these multiple scattering and internal reflection events can increase the overall absorptivity for a given thickness of wafer. In this case, the radiation field inside the sample may resemble that of a blackbody cavity for photon energies that are weakly absorbed. If a beam of light of cross-sectional area A is incident on the front interface of a textured absorber of thickness t, the transmission coefficient, T_1, for a single passage of light is given by

$$T_1 = \frac{1}{\pi L_S A}\int_0^{2\pi}\int_0^{\pi/2} L_S A\cos\theta\exp\left(\frac{-\alpha t}{\cos\theta}\right)\sin\theta\, d\theta\, d\varphi \approx (1-2\alpha t), \quad (2.17)$$

where φ is the azimuthal angle and θ is the angle with respect to the surface normal (perpendicular to the surface). The term L is the radiance of the light source. The approximation illustrated in this equation is for the case of weakly absorbed light. In the case of the solar cell, a metal contact serves as a reflector at the back surface. This reflecting back surface forces the light to propagate through the material again, resulting in a transmission of approximately $(1 - 4\alpha t)$ as the light returns to the front (sample-air) interface. At the front interface, a fraction $1/n^2$ escapes the material to be lost, while the rest of the light is reflected (via total internal reflection) to begin the propagation process once more through the wafer. By neglecting surface reflection (which for a perfectly textured wafer is small) the absorptivity for the textured solar cell geometry is given by

$$a(e) = (1 - T_1) \cdot \sum_{i=0}^{\infty} \left[\left(1 - 1/n^2\right) T_1 \right]^i \approx \frac{\alpha}{\frac{1}{4tn^2} + \alpha\left(1 - \frac{1}{n^2}\right)}, \tag{2.18}$$

where the approximation is again for weakly absorbed light. This equation is a more generalized version of that obtained for "randomized" light [1, 13].

The term "randomized" means that the light is scattered within the semiconductor in all directions. In this way, the light within the solar cell resembles the profile in a blackbody cavity or integrating sphere. The related term "textured" refers to a surface of a solar cell that has been etched or shaped so that it is no longer planar. Light incident on the solar cell is forced to traverse the solar cell at defined angles determined by the angles of the facets of the textured surface. Both textured and randomized solar cell surfaces serve to force the light to take long optical paths through the solar cell. The result of this is that wavelengths of light near the bandgap, with a small absorption-coefficient value, are absorbed in the device. In other words, these techniques confine the light within the solar cell so that the absorptivity will be higher.

For solar cells, the absorptivity can be used as a "lumped parameter" to describe the absorption properties of a particular sample of a material. Although it is a constant for a given individual solar cell, the absorptivity is not a constant for a solar cell *material* (e.g., c-Si or a-Si). Using the absorptivity, the maximum expected short-circuit currents, J_{SC}, can be predicted from the integral of the product of the absorptivity and the AM1.5 photon flux given in Fig. 2.2(b). This is useful optical information when accessing a PV technology's merit or in developing fabrication procedures.

An optical aspect not covered in this section is the description of the metal contact grid on the front of the solar cell. Such a grid is shown schematically in Fig. 1.1, and this is the general metallization pattern used in most nonconcentrator solar cells. The optimum metallization pattern involves a trade-off between the light lost due to shadowing by the grid, and increased resistance losses due to the necessity for the charge carriers to travel long distances through the light-absorbing material (e.g., Si) before collection. For further details on grids, the reader is referred to the bibliography for excellent sources on these aspects [1–6].

2.3 Photoluminescence

All materials that absorb light also emit light. If this were not true, absorption of optical energy could result in an increase of absorber temperature above that of the source, in violation of the second law of thermodynamics. There are many types of luminescence: photoluminescence, electroluminescence, and chemiluminescence, to name a few. In electroluminescence, light is emitted upon the application of an electric field (e.g., an LED).

Photoluminescence is the re-emission of light after absorption. A Stokes shift occurs such that the peak wavelength of the absorption is usually shorter than the peak wavelength of emission. This is in contrast to simple reflection, where photon energy and the wavelength of the light are conserved.

There are two types of photoluminescent emission—phosphorescence and fluorescence. Fluorescent light emission stops after the excitation source is no longer incident on the sample. For phosphorescence, the emission decays slowly after the excitation light source is switched off and can often be detected milli-seconds to minutes later. Fluorescence and phosphorescence are terms valid only where spin forbidden transitions can be identified. For inorganic materials, the distinction between these types of "photoluminescence" is often one of time scale, and so the term photoluminescence will be used in this text to describe both.

Photoluminescence is commonly used to study the band structure and defect structure of materials used in the optoelectronics industry and in solar conversion. It has been found that the width of the luminescence spectrum becomes narrower and the intensity increases as the sample temperature is lowered. This allows low-temperature photoluminescence measurements to more easily identify specific defect transitions than measurements taken at higher temperatures. This defect identification leads to an understanding of the recombination and electronic processes involved in the functioning of the device. This approach has gradually resulted in solar cells, made of various materials, with increasingly higher efficiencies. In normal temperature (ambient) solar cell and detector applications, one views photoluminescence as a loss of energy, rather than as a requirement for efficient conversion. In the open-circuit condition, where energy is not extracted, energy dissipation by luminescence is a measure of the device's ability to bypass other losses. These losses compete with the production of work from light (or light from electricity in the case of the LED). The current-voltage characteristics and luminescent radiance for a quantum solar energy conversion system are directly connected. As shall be explored in Chapter 5, the luminescence, while constituting the only unavoidable current loss, can be used to predict the maximum voltage expected from a solar cell material. By observing the luminescence, one can therefore assess the quality of the absorber material and improve device-fabrication techniques that may lead to increased conversion efficiencies. Here, the concept of luminescence as an optical property of all solar cells is introduced, and the basics of the optical measurements that can quantify luminescence in solar cells are described.

An example of an experimental setup used to quantify photoluminescence is shown in Fig. 2.11. Light excites the sample, and the emitted radiation is collected and analyzed. Figure 2.12 shows a typical output photoluminescence spectrum for a solar cell grade Si wafer induced using a continuous-wave (CW) argon-ion laser (at 488 nm) and recorded using a spectrometer. A typical commercial instrument consists of a visible and an IR single-grating monochromator, a 610-nm-long wavelength-pass filter, a detector-amplifier package, and a fiber-optic-based optical coupling. A thermoelectrically cooled InGaAs photodiode can record the IR signal, while a multi-alkali photomultiplier is often used for the visible wavelength range. The argon-laser power used to take the measurement in Fig. 2.12 was 40 mW, the step size of the instrument was 10 nm, and the average scan time was 200 seconds. The laser irradiance was adjusted so that the absorbed number of photons was similar to what would be obtained under AM1.5 sunlight. As can be seen, the photoluminescence peak is near the bandgap wavelength and can be used to estimate this solar cell parameter. The photoluminescence efficiency, Φ, i.e., the number of photons

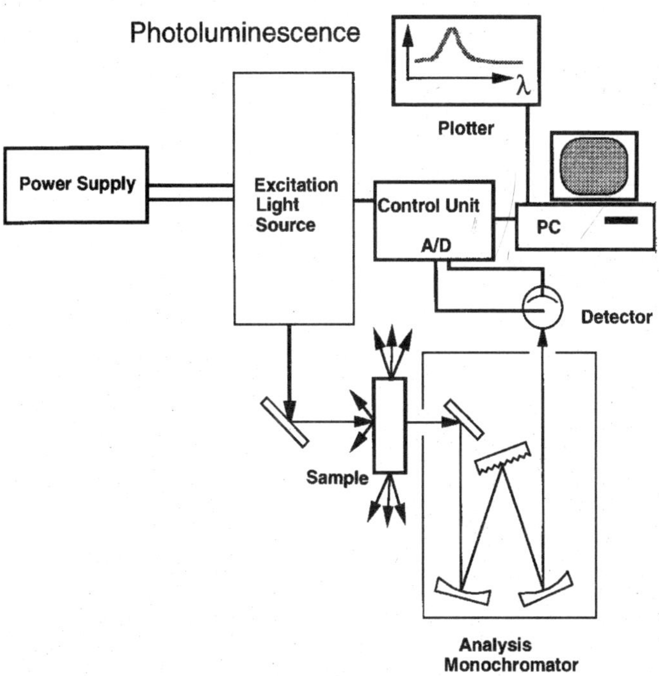

Fig. 2.11 Experimental setup used to determine the photoluminescence of a photoconverter material. Excitation via a laser induces photoluminescence that is coupled into the analysis monochromator. Calibration of the signal can be performed using a sample of known photoluminescence efficiency.

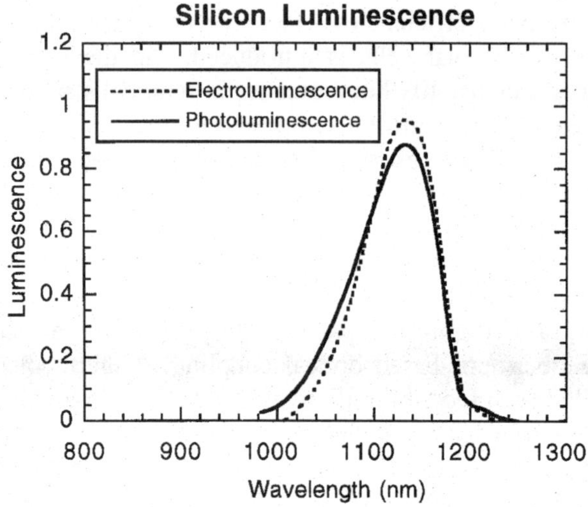

Fig. 2.12 Room temperature silicon photoluminescence (solid) and electroluminescence (dashed). The electroluminescent silicon LED was driven at 0.65 V and 50-55 mA. Both spectra were measured with an Instrument Systems GmbH Spectro 320 portable spectrometer using the same y-axis scale.

output per input photon, for the Si wafer is obtained from the integral of the curve in Fig. 2.12. Here, the photoluminescence efficiency is found to be between 10^{-4} and 10^{-5}. As a comparison, the upper limit of Φ values observed for pure Si wafers that can be used to fabricate state-of-the-art Si solar cells is 10^{-3}.

If a solar cell is indeed constructed with the wafer used for the above measurement, the open-circuit (or maximum) voltage measured in the laser or under AM1.5 illumination was 0.62 V. This value can validate the usefulness of photoluminescence measurements on materials used for solar cells if it is compared to the electroluminescence from the same device.

To measure the electroluminescence, a solar cell can be mounted in the same machine as for photoluminescence and a voltage applied to the sample without the use of the external exciting light. In this example, the voltage applied was 0.62 V, and represents the driving of the device as a Si LED. At an applied current of approximately 50 mA, the device had an electroluminescence efficiency between 10^{-4} and 10^{-5}. The Si LED has a low efficiency, but this is not the point. Since the magnitude and shape of both the electroluminescence and photoluminescence match, one can easily see that the photoluminescence measurement on a wafer destined for a solar cell can be used to predict the open-circuit voltage expected on the completed devicc. Therefore, photoluminescence is used as a quality control tool to reject materials or different samples of the same material that are not expected to perform well. It can also be used to screen new materials that are being considered for novel types of solar cells.

Figure 2.13 shows an example of a time-resolved photoluminescence measurement that can be used to extract one of the vital parameters in solar cell

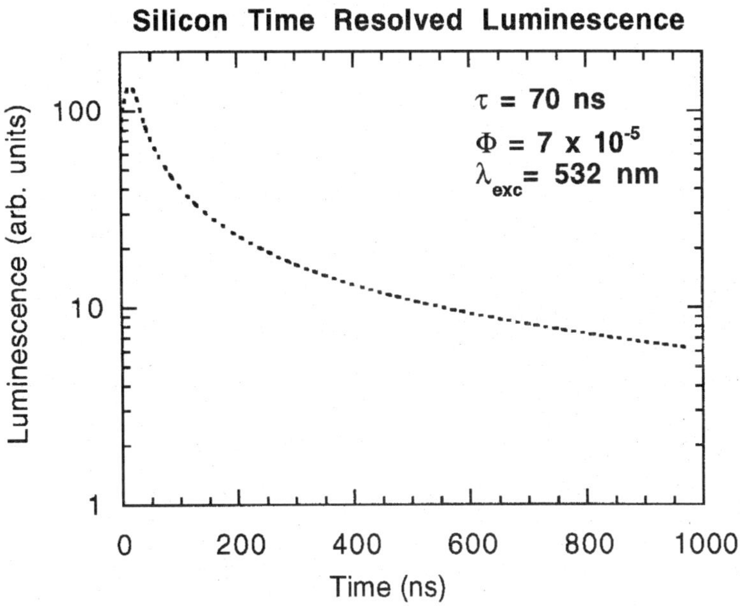

Fig. 2.13 Measured time-resolved photoluminescence for a passivated (SiO_2-coated) silicon wafer.

design—the lifetime of the excited state, τ. Instead of being continuously illuminated, as in the photoluminescence experiment described above, a brief laser pulse (e.g., at 532 nm) is applied to a sample (for Fig. 2.13, a Si wafer)where electrons and holes are produced, and then rapidly recombined. Some of these electrons recombine radiatively, and this photoluminescence can be detected and plotted as a function of time (i.e., from the end of the laser pulse.) The lifetime is defined as the 1/e fall-off time for photo-produced electrons. For the example in Fig. 2.13, it is 70 ns. This parameter is strongly affected by the quality of the solar-absorber material. Recall that charge carriers must have a sufficient lifetime to diffuse and drift to the external contacts to be collected. If recombination is too swift, few carriers will exit the solar cell and low solar-conversion efficiency will result. The lifetime is therefore a performance parameter for solar cells and solar cell materials. This is discussed in Chapter 3.

Photoluminescence is not the only way to measure the lifetime of the excited state. Several techniques are used. It is worth mentioning the time-resolved microwave conductivity measurement (TRMC) in the optical context. Similar to the photoluminescence measurement described above, a laser pulse can excite electrons into the conduction band of the material. Instead of measuring the luminescent light that is emitted by the sample, a microwave beam can probe the sample and monitor the (microwave) absorption by free electrons. Figure 2.14 shows an example of such a measurement, and it demonstrates that the type and

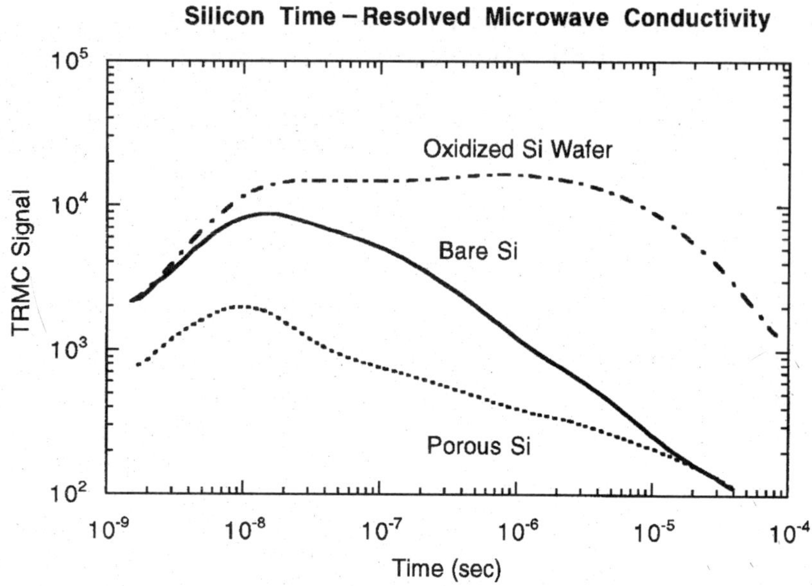

Fig. 2.14 Double logarithmic representation of the time-resolved microwave detected photoconductivity (TRMC) for a silicon wafer with three different coatings. The porous Si-coated Si wafer (bottom dashed line), is produced by electrochemically (anodic) etching the wafer in 20% acid (HF) in ethanol for 10 minutes at 20 mA/cm^2. The bare Si (middle line) was produced by etching any oxide off the wafer with HF. Thermally oxidized Si wafer (top line) represents the original Si wafer used for the other two conditions. The oxide was produced by heating the wafer in air at 600°C. For the TRMC measurement, the excitation wavelength was 532 nm at an excitation density of 0.5 μJ/cm^2.

condition of the sample can be determined in this manner. Time-resolved measurements are strongly influenced by the presence of traps and recombination centers (RC) shown in Fig. 1.4. In the next section, the solar cell parameters introduced in this chapter shall be used to delve into more detail as to how solar cells function and how to view their electrical characteristics from an optical viewpoint.

3

Solar Cell Equations

3.1 PV Device Characteristics

Figure 3.1 illustrates, again, the basic operation of a solar cell showing excitation by light, charge separation, migration, and collection. The measurement of a solar cell's current-voltage curve is of prime importance in solar cell characterization. It represents the electrical output characteristics that are the result of the processes shown in Fig. 3.1. As stated in Chapter 1, the I-V characteristic is described by

$$I(V) = I_{SC} - I_0\left[\exp\left(\frac{qV}{\gamma kT}\right) - 1\right]. \qquad [3.1(a)]$$

The term γ in the denominator of the exponent is called the diode-quality factor, or the "ideality" factor. For a perfect diode, γ is unity, but its value ranges from 1 to 2 in typical devices. The multiplier I_0 is called the "saturation current" and is

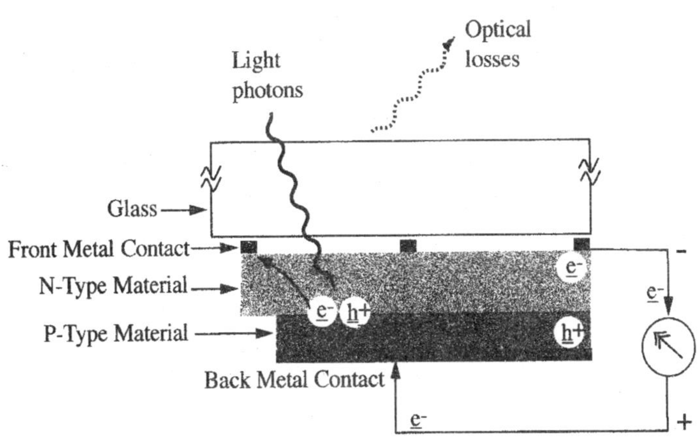

Fig. 3.1 Side view schematic of a solar cell showing the various layers, the process of charge separation charge transport and, finally, charge collection by the external contacts.

the expected current at reverse bias in the dark. The right-hand side of this equation starting at the subtracted term is called the "diode equation" used to describe the electrical characteristics of a rectifier. For nonideal devices, several terms similar to the form given in the diode equation are often used to describe the output in different voltage regions. Each of these terms uses separate values for I_0 and γ for each range of voltages.

From the above equation, and solving for the open-circuit voltage (I = 0), one obtains

$$V_{OC} = \frac{\gamma kT}{q} \ln\left(\frac{I_{SC}}{I_0} + 1\right). \quad [3.1(b)]$$

It can be seen that the open-circuit voltage is a function of the dark and light currents as well as the diode-quality factor. The output characteristics are often expressed as current density (e.g., in mA/cm^2), taking into consideration the area of the device, A. In this case, $J = I/A$ and $J_0 = I_0/A$. The use of current density, rather than just the current, is therefore useful in comparing different devices or different materials.

Figure 3.2 demonstrates yet another way of looking at a solar cell—the electrical model. The input of solar energy is shown as a current generator, I_{SC}.

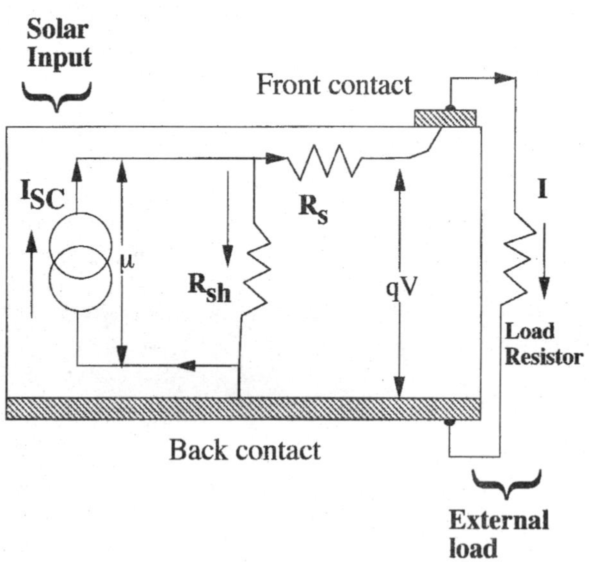

Fig. 3.2 Circuit diagram for a solar converter illustrating the relationship between the maximum voltage, chemical potential, μ, series resistance, R_s, and voltage appearing across the load resistor, V. Radiative and nonradiative recombination losses are indicated as resistors which shunt the current source, I_{SC}.

Recombination is shown as a shunt resistor, and a series resistor blocks the passage of current I to the external load via the external contacts. Instead of voltage alone, the term qV is used in this diagram. It is dimensionally equal to the energy per particle, and the unit most convenient for solar cells is the electron volt, eV/particle (e.g., eV/electron). Alternatively, Joules/particle can also be used. The term μ is not often used in photovoltaic descriptions, but is quite common in chemistry. It is the chemical potential or energy per particle, or it can alternatively be expressed as the energy per mole. It is also called the Gibbs Free Energy. In the case of solar cells discussed here, it represents the difference (at opposite sides of the cell) between the energy per electron and the energy per "hole." This chemical potential represents the maximum driving force that can be generated in the solar cell in the absence of losses, such as shunt and series resistance. This aspect shall be described further in Chapter 5 with regard to luminescence and the optical properties of a generalized quantum solar converter.

The I-V equation can also be further modified to include resistive losses due to series and shunt resistance shown in Fig. 3.2, thus Eq. [3.1(a)] becomes

$$I(V) = I_{SC} - I_0\left[\exp\left(\frac{qV - IR_S}{\gamma kT}\right) - 1\right] - \frac{(V - IR_S)}{R_{sh}}. \qquad [3.1(c)]$$

Series resistance, R_S, is due to the conductivity of the materials and the thickness of the various layers. Shunt resistance, R_{sh}, is due to the short circuit pathways that allow charge carriers to recombine before they can be collected at the contacts and forced to do work in an external circuit. These include electrical shorts, as well as nonradiative recombination depicted in the energy band diagram of Fig. 1.4.

A simplified I-V measurement setup is shown in Fig. 3.3. Shown in the upper diagram, the voltmeter measures the voltage across the converter while the current is simultaneously measured by a current meter (or electrometer). Shown in the lower diagram, a simple variable load (resistor) may be used to record the power-producing portion of the I-V curve. For example, for solar cells that have an output voltage in the range 0.1–0.7 V and currents in the range 0.5–30 mA, a 500-Ω potentiometer can be employed. Of the two types of setups for I-V curve determination, the use of a simple variable load more closely resembles the conditions found during practical solar converter operation. In addition, it has the least chance of damaging the converter, which may be sensitive to large reverse or forward voltages. Potentiostats and current-voltage profiling equipment used in the semiconductor industry can also be used so as not to subject the solar cell to excessive voltages or currents.

As an example, I-V curves for a Si solar cell can be taken using an Oriel-AM1.5 solar simulator, which produces an irradiance of 1000 W/m^2 in an area 100 × 100 mm^2. Figure 3.4 shows the J-V curve for the dark, and for the solar

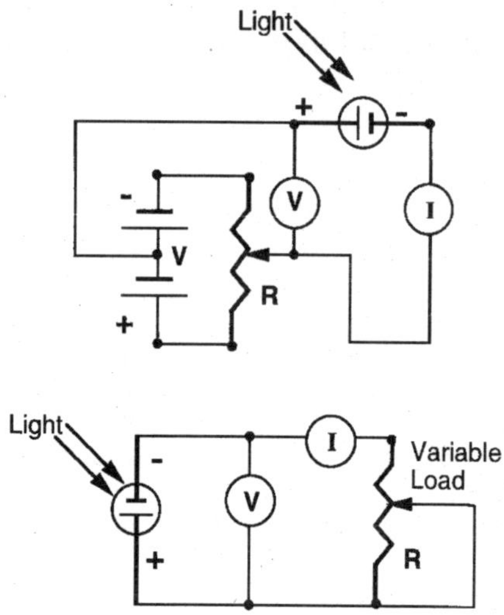

Fig. 3.3 Two possible choices of electrical connections for the measurement of the current voltage, J-V, characteristics of photoconverters that produce an electrical output. Shown at top is the configuration using a variable applied voltage. At bottom is shown the variable load configuration used for sensitive photoconverter devices.

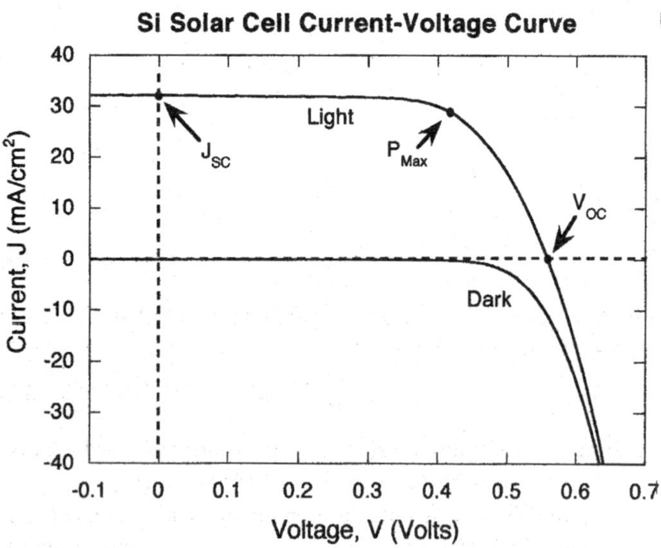

Fig. 3.4 Typical Si solar cell current voltage (J-V) characteristics showing the open circuit voltage, short circuit current and maximum power point. The bottom curve is the current passed through the device in the dark. As can be seen, the PV device in the dark acts as a diode, passing current easily in only one direction.

cell illuminated by AM1.5 simulated sunlight. The solar cell tested employed a P-type base layer, the same as that used in the optical studies described in the previous chapter. The current-voltage curve of a solar cell yields important operational parameters that are dependent on the terms in Eq. [3.1(c)], among which are the short-circuit current I_{SC}, the open-circuit voltage V_{OC}, the current, I_{mp}, and voltage, V_{mp}, at the maximum power point, P_{max}. The term called the fill factor, or FF, combines these terms and is defined as

$$FF = \frac{P_{max}}{V_{OC}I_{SC}} = \frac{V_{mp}I_{mp}}{V_{OC}I_{SC}}. \tag{3.2}$$

Note that the fill factor does not depend on the area of the device. For practical Si photovoltaic cells, the fill factor ranges from approximately 0.70 to 0.85. For a typical single-crystal silicon solar cell, we find that V_{OC} = 0.5 to 0.6 V, J_{SC} = 35 mA/cm^2, and FF = 0.8. Thus, a 100-cm^2 cell illuminated with 100-mW/cm^2 sunlight is capable of producing approximately 1.6 W (e.g., 16% efficiency). Most practical applications would require more power. Fortunately, one can increase the voltage by adding more cells in series, and one can increase the current by wiring more cells in parallel. This is how PV modules are produced.

The solar conversion efficiency is given by

$$\eta = \frac{FF \cdot V_{OC}J_{SC}}{P_S}, \qquad [3.3(a)]$$

where P_S is the input solar irradiance (e.g., in W/m^2 or mW/cm^2). This quantity is best measured with a broadband pyranometer (sensitive to all wavelengths), or using a calibrated detector with a similar spectral sensitivity as the device (solar cell) under test.

3.2 Quantum Efficiency for Current Collection

If all the photons that were absorbed in a solar cell could be collected by the external contacts, it would be no trouble to predict the short-circuit density (J_{SC}) values from the absorptivity, as described in Chapter 2. Unfortunately, not all electron-hole pairs created via the absorption of light are collected at the external contacts. While the "induced" current in the cell is approximately equal to the number of photons absorbed within a material's bandgap range, some of the induced charge carriers recombine before they can diffuse and drift to the external contacts where they are collected. The incident photon-to-current efficiency is the product of the quantum absorptivity and the current collection quantum efficiency, η_c. It is given by the ratio of the number of electrons that make it out of the solar cell to the number of incident photons. For a typical solar cell, the ratio is given by

$$\eta_q = \frac{\text{number of electrons flowing through external circuit}}{\text{number of incident photons}} = a(e)\cdot\eta_c \ . \qquad [3.3(b)]$$

This photocurrent efficiency is sometimes also called the incident photocurrent efficiency (IPCE) or the external quantum efficiency. The J_{SC} value for a typical solar cell can be predicted from the integral of the product of the absorptivity, the current-collection quantum efficiency, and the AM1.5 photon flux displayed in Fig. 2.2(b). For some solar cells, additional factors must be included on the right-hand side of this equation. For example, the dye-sensitized solar cell (to be discussed in more detail later) must include an "injection" efficiency. This describes the probability of the electron to be transferred to another material after it has been excited by light. As another example, one can consider the factors that determine η_q for organic solar cells that use organic-dye molecules, or polymers, as the light-absorbing material. For these devices, an efficiency term must be included that describes the charge-separation efficiency of "bound" excited carriers such as excitons or polarons. The general principle here is that the sequence of events and processes that lead from light absorption to charge collection by the contacts must be considered in any theoretical description of the quantum efficiency. Of course, the measurement of the quantum efficiency is comparatively straightforward.

A typical measurement apparatus for this type of measurement is shown in Fig. 3.5. Monochromatic light illuminates the solar cell device under test. The cell is typically biased to the short-circuit condition. A white light "bias" is also added to simulate the response under conditions that approximate AM1.5. This is because, ideally, one wishes to determine the incremental change in photocurrent at a given wavelength when the radiance is close to that of sunlight. Some solar cells are nonlinear and respond differently to low-intensity light than they do to a high-intensity beam. If the intensity from the monochromator is sufficient, the white light bias, lock-in amplifier, and chopper can be omitted.

An example of the external quantum-efficiency measurement for Si, together with the AM1.5 spectrum, is shown in Fig. 3.6 for comparison. As can be seen, Si absorbs well throughout the solar spectrum, and devices that can collect most of the absorbed light are within current technological capabilities. Operationally, the quantum efficiency is calculated from a measurement of the short-circuit current at a given wavelength, taking into consideration the input radiant power. Considering the various conversion factors between the area, the current, and a number of electrons, the quantum efficiency is given in a useful form by

$$\eta_q = \text{IPCE}\ [\%] = \frac{1239 \times J_\lambda\ [\mu\text{A/cm}^2]}{\lambda\,[\text{nm}] \times P_\lambda\ [\text{W/m}^2]} \ . \qquad [3.3(c)]$$

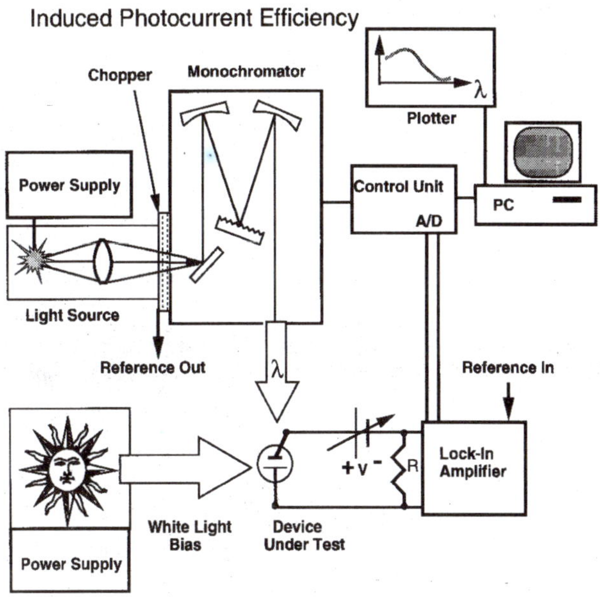

Fig. 3.5 Typical setup for the determination of solar cell photocurrent spectra, and quantum efficiency otherwise known as IPCE. Chopped monochromatic light produces a photocurrent through and a voltage across the resistor. The voltage is read by the lock-in amplifier referenced to the chopper frequency. A continuous white light bias is applied to the cell to allow for operation near solar illumination levels. A variable voltage source, V, is used to bias the device so that operation is near the short circuit condition. The output of the PV solar cell is recorded as the wavelength is varied.

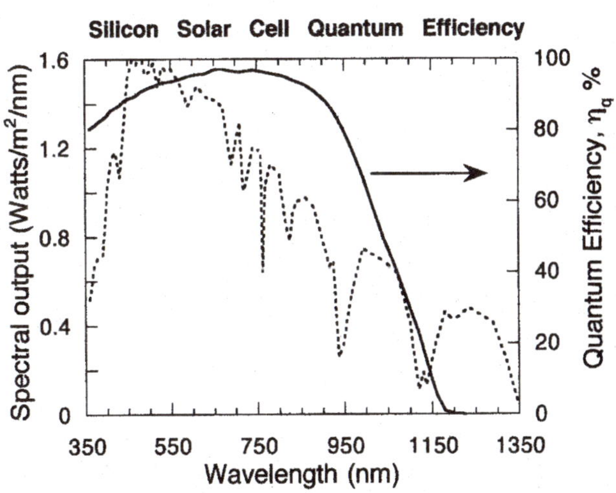

Fig. 3.6 Measured Si solar cell IPCE overlaid on the AM1.5 solar spectrum (dashed line).

The units corresponding to the use of the 1239 conversion factor are given in brackets. It should be noted that P_λ is often measured using a calibrated detector, or calibrated solar cell. The units of calibration are typically in A/W. In this case, the reference detector's current is measured and converted into an equivalent W/m^2 value before use in the above equation.

3.3 Lifetime, Diffusion Length, and Electron Concentrations

For the standard solar cell geometry used in most single-crystal and thin-film solar cells, the current-collection quantum efficiency, which can be used to predict the measured J_{SC}, can itself be predicted from basic optics and semiconductor physics. To understand how to derive an equation for the current-collection quantum efficiency, one can solve an equation representing the diffusion, recombination, and generation of minority carriers within the light-absorbing region of the solar cell. The following analysis is an illustrative simplification taken from the case of a P-type base region of a P-N-junction cell. It is nonetheless applicable for any solar converter where generation of charge carriers (by light) is followed by transport, recombination, and diffusion. Using the standard-transport-equation analysis, the electron distribution and the spatial dependence of the Fermi levels can also be estimated. In this conventional approach, Fermi statistics can be used to describe the particular quantum converter represented by a P-N junction solar cell. It is first instructive to introduce the terms used in standard solar cell-device physics.

First, we can define L as the diffusion length (e.g., for electrons), and recall that α is the absorption coefficient of the light absorber (e.g., for Si). The diffusion length is a measure of the distance a charge carrier can diffuse during its lifetime. The relationship between the diffusion length and the lifetime of the excited state is

$$L = (D\tau)^{1/2}. \quad (3.4)$$

Here, D is the diffusion constant given by Einstein's relation,

$$D = \frac{kT\mu}{q}, \quad (3.5)$$

where k is Boltzmann's constant, q is the elemental charge, T is the absolute temperature expressed in kelvins, and μ is the mobility of the charge carrier. For example, this applies to a τ value of 200 ps (10^{-15} seconds), $D = 9 \times 10^{-6}$ cm^2/s, and μ is 3.5×10^{-4} cm^2/V·s. Typical values for Si wafers used in solar cells are between 10 and 100 cm^2/V·s. Note that the symbol "L" is used in this text (and others) for both radiance in the optical context and the diffusion length for the transport of charges. The symbol μ is used for chemical potential in Chapter 5 and for charge-carrier mobility in the analysis that follows.

One can determine how many charge carriers (n_0, p_0) are present in the dark within a solar cell, and how many are produced via irradiation by light (n^*, p^*). A useful relationship to determine the concentration of electrons and holes is

$$n_i^2 = n_0 p_0 < n^* p^*, \tag{3.6}$$

where n_i is called the intrinsic concentration and the subscript 0 applies to the equilibrium condition. This equation is also called the law of mass action. Note that this expression applies to the N-type or P-type region. In a doped semiconductor, electron acceptors are abundant in the P-type side, and electron donor molecules are added to the N-type side. The ionization of these dopants creates additional free carriers in the solar cell. For charge neutrality, the following relationship holds true for both P-type and N-type materials:

$$n_0 + N_A = p_0 + N_D, \tag{3.7}$$

where N_A is the concentration of ionized acceptors and N_D is the concentration of ionized donors. In most practical cases, the total dopant concentration is approximately equal to the ionized dopant concentration (in other words, all dopant atoms are ionized). As an example, for the P-type region, only excess acceptors are present, and their concentration is much larger than the value of n_0. Therefore, $p_0 = N_A$ in the dark, and from Eq. (3.6) we obtain

$$n_p = \frac{n_i^2}{N_A}. \qquad [3.8(a)]$$

For example, an acceptor concentration of $N_A = 1 \times 10^{17}$ cm^{-3}, and $n_i = 1.1 \times 10^{10}$ cm^{-3} for silicon, gives us $n_p = 1.2 \times 10^3$ cm^{-3}, the dark concentration of electrons. One can compare this electron-concentration value with the concentration of excited electrons when a solar cell is biased. For this case, the electron concentration, as a function of its position in the device, is given by the difference in Fermi levels, ($E_{fn} - E_{fp}$). The concentration of electrons as a function of distance, x, is given by

$$n_p(x) = \frac{n_i^2}{N_A} \exp\frac{(E_{fn} - E_{fp})}{kT_0}. \qquad [3.8(b)]$$

Referring to Fig. 1.3 (Chapter 1), one sees that V is obtained from the difference in Fermi levels at the external contacts. With this introduction of terms, a calculation of the photocurrent and quantum efficiency in a solar cell can be made from a solution of the "transport equation."

3.4 The Transport Equation and Current Extraction

Imagine a small volume of material in the quasi-neutral region (i.e., P-type base), or absorber layer, of a solar cell. In the limit of negligible electrical fields, the one-dimensional equation describing the fate of electrons in this volume is given by the diffusion of these charge carriers, and the difference between the generation (via light) and recombination of charge carriers. Mathematically, this is expressed by the transport equation

$$-\frac{1}{q}\cdot\frac{dJ}{dx} = G - U, \qquad [3.9(a)]$$

where J is the current density (i.e., mA/cm^2), and G is the rate of generation of charge carriers produced by the absorption of light. The recombination rate, U, is the ratio of the "excess" electron concentration to the lifetime of the electron.

The current density, J, is dependent on the change in the carrier concentration as a function of position, x, within the solar cell, or

$$J = -qD\frac{dn_p(x)}{dx}. \qquad (3.10)$$

Therefore, Eq. [3.9(a)], in its simplest form, becomes

$$D\frac{d^2 n_p}{dx^2} - \frac{(n_p - n_{p0})}{\tau} = -\alpha\Gamma_{S\lambda}\exp(-\alpha y), \qquad [3.9(b)]$$

where the incident photon flux is $\Gamma_{S\lambda}$, D is the diffusion coefficient, and n_p is the electron concentration in the P-type base layer of the solar cell. We let $y = (x - x_p)$, which is the distance from the edge of the junction to the volume element in question. The quantity x_p is the position of the junction (the depletion-layer edge). The value of n_p in the dark (an equilibrium value) is n_{p0}. The first term in the equation represents diffusion. The second term represents both radiative *and* nonradiative recombination, and the third term represents the generation of electrons by light. This term is similar to that of the Beer-Lambert equation of Eq. [2.8(a)], except that a correction accounting for the amount of light absorbed before light reaches a given volume element is made. This neglects any absorption by the thin N-type layer of the solar cell that is in front of the P-type layer being considered in this analysis. If the absorber (wafer) thickness is t, then the boundary conditions for the above equation are (1) the electron concentration approaches its equilibrium (dark) value at the back surface, $n_p(t) = n_{p0}$, and (2) $n_p(0) = n_{p0}\exp(qV/kT_0)$ at the junction ($x = x_p$, and $y = 0$). From Eqs. [3.8(b)] and [3.9(b)], one determines that the recombination term used in standard P-N junction theory is exponentially dependent on the voltage. This is a common assumption in solar cell models, and it implies that the nonradiative recombination rate is a constant fraction of the radiative recombination. Both radiative and nonradiative recombination are assumed to depend exponentially on the voltage. This assumption will also be applied in Chapter 5.

The above equation allows one to determine the concentration of electrons in the device as a function of distance. These electrons can then be counted and summed to find the current-collection quantum efficiency described previously. Using the boundary conditions given above, the solution of the transport equation is

$$n_p(x) = \frac{n_i^2}{N_A}\left[\exp\frac{qV}{kT_0} - 1\right]\exp\left[\frac{-y}{L}\right] + \frac{\alpha\Gamma_{S\lambda}}{D(\alpha^2 - 1/L^2)}\left[\exp\frac{-y}{L} - \exp(-\alpha y)\right] + \frac{n_i^2}{N_A}, \tag{3.11}$$

where the term y is equal to $(x - x_p)$. The absorption coefficient is given by α, $\alpha = \alpha(\lambda)$, and the diffusion length, L, is related to the charge carrier mobility, μ, and carrier lifetime, τ. The carrier lifetime can be deduced from time-resolved luminescence measurements as previously described. The term n_i is the intrinsic carrier concentration and the doping concentration is given by N_A. A typical plot of the electron concentration is shown in Fig. 3.7(a). The difference in Fermi levels is displayed in Fig. 3.7(b) and overlaid on the band diagram of the solar cell in Fig. 3.8.

The photocurrent at a given wavelength is obtained from the above equations because all extracted current must pass through the junction at $x = x_p$. The photocurrent at a particular wavelength is then given by the solution of Fick's law [Eq. (3.10)] at this position:

$$J_\lambda = J(x \to x_p) = \left[-qD\frac{dn_p(x)}{dx}\right]_{x=x_p}. \qquad [3.12(a)]$$

The solution of this equation yields an equation for J_λ as complex as Eq. (3.11). The resulting equation, however, is of the form

$$J_\lambda(V) = J_{\lambda SC} - J_0\left[\exp\left(\frac{qV}{kT}\right) - 1\right], \qquad [3.12(b)]$$

which, again, yields the diode equation [Eq. (3.1)] if the current generated at each wavelength is summed. The current is said to obey the law of superposition if the light-generated current is superimposed on the value of the current in the dark. This is true for some solar cells, but, in general, it may not always be the case.

The current-collection quantum efficiency at a particular wavelength is given by Eq. [3.12(b)] and Eq. [3.3(b)] in the limit of zero applied voltage (i.e., short-circuit conditions). Using the simplifying assumption of an absorber-layer thickness (P-layer) that is much larger than the diffusion length, L, one obtains

$$\frac{J_{\lambda}/q\Gamma_{S\lambda}}{a(\lambda)} = \eta_c \approx \frac{1}{1+\dfrac{1}{\alpha(\lambda)\, L}}, \tag{3.13}$$

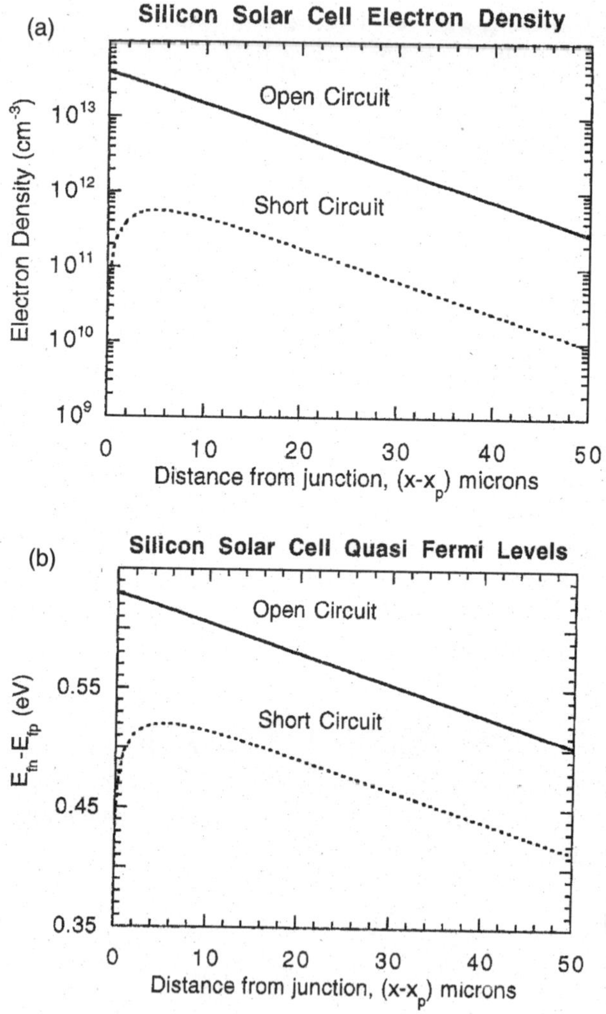

Fig. 3.7 (a) Calculated electron density at open and short circuit for a silicon cell using the transport equation, AM1.5 illumination and typical device values of $N_A = 1 \times 10^{17}$ cm^{-3}, L = 10 μm, D = 25 cm^2/s, α = 3333 cm^{-1} and n_i = 1.10 x 10^{10} cm^{-3}, (b) Chemical potential (difference between the quasi Fermi levels) for the conditions in (a). The plots are the result of Eqs. [3.8(b)] and (3.11), and do not include the effects of surface recombination.

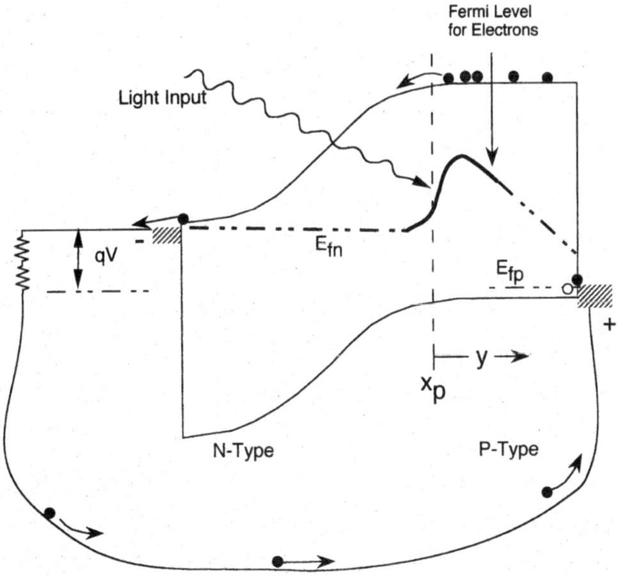

Fig. 3.8 Solar cell band diagram overlaid with the shape of the profile in Fig. [3.7(b)].

where $\Gamma_{S\lambda}$ is the photon-flux density at a given wavelength. This equation is used extensively for P-N-junction solar cells, for Schottky-barrier solar cells, and for photoelectrochemcial cells. For example, given a quantum-efficiency value of near 0.9 and an absorption-coefficient value of $10^3 cm^{-1}$, the L value required for efficient charge carrier collection is approximately 90 μm. The typical mobility that corresponds to this value is between 10 and 100 $cm^2/V \cdot s$.

In general, the collection efficiency can be predicted from the solution of Eq. [3.12(b)] and then compared to measurements such as the one shown in Fig. 3.6. The maximum photocurrent output for a solar cell is given by the integral (over photon energy, or over wavelength) of the product of the AM1.5 photon flux, absorptivity, and the collection efficiency outlined above. We have seen from the preceeding analysis that the use of basic optical and material properties allows insights into charge-carrier collection and the spatial dependence of the electron concentration and Fermi levels in a solar cell.

4
Photoelectrochemistry

4.1 Basic Photoelectrochemistry

There are two types of solar cells—those in a solid-state form and those containing liquids. The latter is often called a photoelectrochemical solar cell. Figure 4.1 shows a simplified diagram of a photoelectrochemical, or PEC, solar cell. The top of the figure shows the equilibrium-energy-band diagram for a solar cell material (without a P-N junction) immersed in a solution containing a redox (oxidation-reduction) active species. For example, a Si wafer might be immersed in an acetonitrile-based solution containing iodide as a redox active species. Analogous to the case of the P-N junction described in Chapter 1, the exchange of charge carriers can occur between the redox species and the semiconductor (e.g., N-type Si) to create a depletion region. In the light, electrons can be excited to the conduction band, just like in the case of the P-N junction, and can be collected at the back contact. The role of the redox species, sometimes called the redox mediator, is to replenish the electrons as they are lost and to complete the electrical circuit via a counter electrode. This electrode is usually coated with a suitable catalyst (e.g., C, Pt or Pd) that facilitates the regeneration of the redox species. The redox species therefore goes through a cycle of oxidation and reduction. As shown in Figs. 4.1(a) and 4.1(b), a PEC solar cell operates like one half of a P-N junction. In this regard, it resembles a Schottky barrier solar cell. For both of these types of cells, the equations in Chapters 2 and 3 apply. Practical applications of PEC solar cells have been limited by stability and corrosion issues. If the semiconductor does not absorb light very well itself, a "sensitizer" can be attached to the surface to act as the light absorber. This is the basis of the dye-sensitized solar cell shown at the bottom of Fig. 4.1. This type of photoelectrochemical cell has come the closest to reaching practical efficiencies and commercial applications. The most efficient (and promising) dye-sensitized solar cell utilizes a semiconductor that is nanocrystalline.

4.2 The Dye-sensitized Nanocrystalline Solar Cell

The dye-sensitized nanocrystalline solar cell is remarkable in that it resembles natural photosynthesis in two respects: (1) it uses an organic dye to absorb light

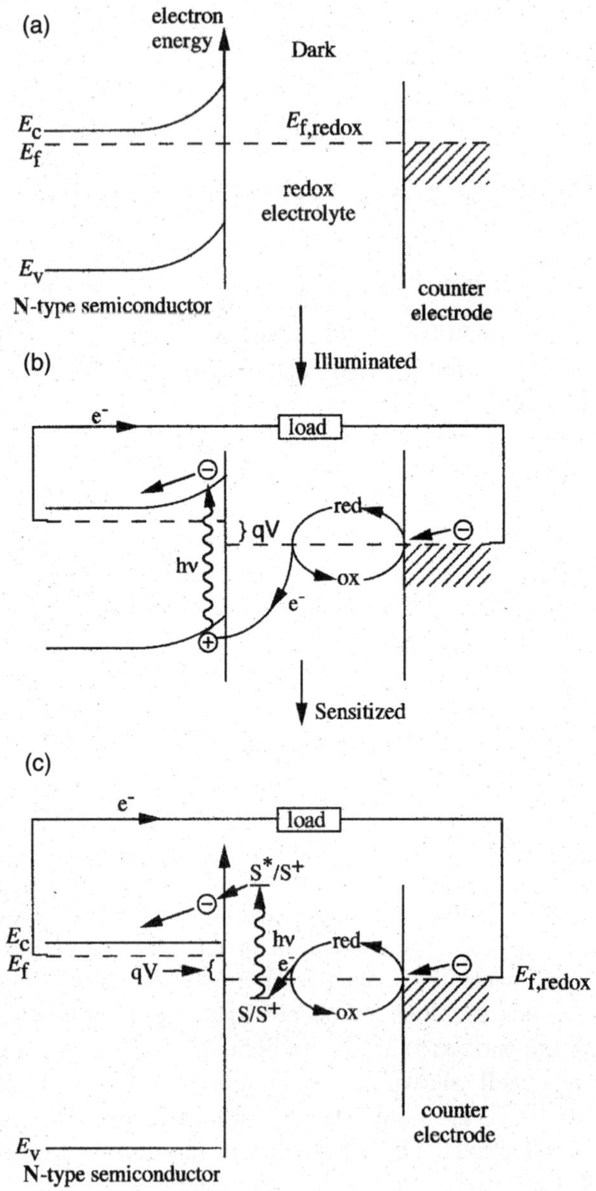

Fig. 4.1 An energy band diagram for regenerative photoelectrochemical solar cells. a) An N-type semiconductor immersed in an electrolyte in the dark, b) the same device in the light, and c) in the light when a sensitizer is adsorbed to the semiconductor. The last photoelectrochemical device represents the dye-sensitized solar cell (DSSC). E_c is the energy level of the conduction band, E_v is the energy level of the valence band, and red and ox are the reduced and oxidized states of the redox electrolyte. (Diagram courtesy of Andreas Kay.)

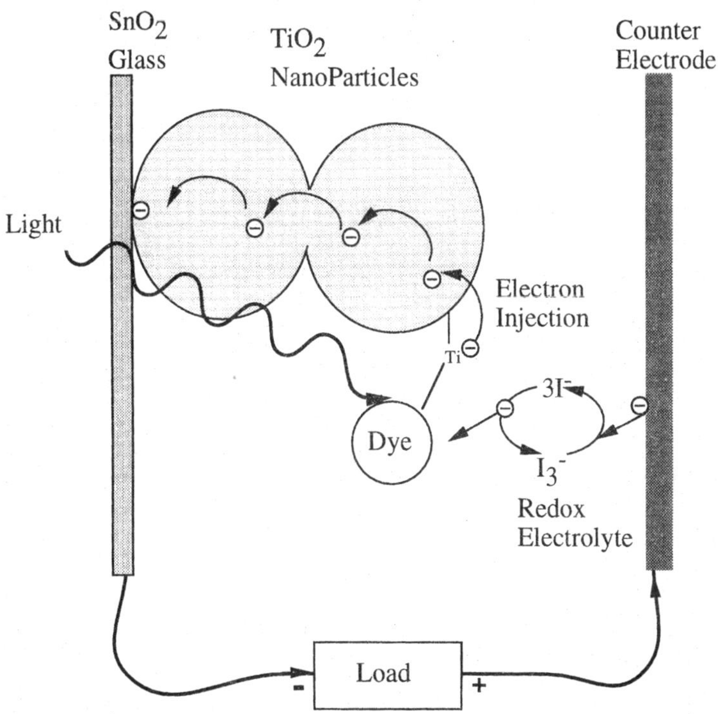

Fig. 4.2 Operation of the dye-sensitized nanocrystalline TiO_2 solar cell. Light is absorbed by the dye molecule (sensitizer) and the resulting excited electron is "injected" into the TiO_2. The electrons then diffuse within the porous TiO_2 structure and are collected at the back contact (conductive transparent glass). The resulting positive charge on the dye is compensated by the mediator, which itself is reduced after the electron has passed through the external load. The redox mediator in the electrolyte goes through cyclic oxidation and reduction as the process continues.

and produce a flow of electrons, and (2) it uses multiple layers to enhance both the absorption and collection efficiencies of previous approaches. This solar cell device is of a new class called molecular electronic devices, otherwise known as nanotechnology. A schematic for this device is shown in Fig. 4.2.

In the late 19th century, it was discovered that certain organic dyes could extend the response of silver-halide-based photographic film to visible wavelengths. The mechanism has been found to involve the electron, or energy transfer, from the organic molecule to the semiconducting silver-halide grain. This sensitization is the basis for modern photography. The dye-sensitized nanocrystalline cell uses the same type of sensitization to create an efficient, but not yet commercially available, solar cell. This device is so simple that it can be made in a matter of a few minutes in a "kitchen laboratory." [16]

To create this newer-generation solar cell, a solution of titanium dioxide in nanometer-size particles, or TiO_2, is deposited directly on conductive glass by a process similar to that used in painting. In fact, TiO_2 powder is one of the cheapest known large-bandgap semiconductors, is abundant, and is currently used in white paints. The film is heated to form a porous, high-surface-area TiO_2 structure that resembles a thin sponge or membrane. This is used as a support while the glass plate is dipped into a solution such as a red ruthenium containing dye, or, alternatively, an anthocyanin, or green chlorophyll derivative. A single layer of dye molecules attaches (by adsorption) to each particle of the TiO_2, and acts as the primary absorber of sunlight. To form the final cell, a drop of liquid electrolyte containing iodide, I^-, is placed on the film to percolate into the pores of the membrane. A counter electrode of conductive glass, coated with a thin catalytic layer of platinum or carbon, is placed on top, and the resulting sandwich is illuminated through the TiO_2 side.

Relatively thick layers of organic dyes have been used previously for what are called organic solar cells. Since the mobility of charge transported within the organic layer is so low, it was found that only a very thin layer was active and necessary for charge injection. This meant that if thicker layers were used to absorb more light, they would not proportionally add to the electrical output of the cell. What is new about the nanocrystalline cell is the use of a rough TiO_2 substrate acting with a thin layer of dye in order to increase the light absorption while allowing for efficient charge collection. Since the layer of dye is so thin, the excited electrons produced from light absorption can be injected into the TiO_2 with near-unity efficiency via sensitization. The TiO_2, therefore, functions in a similar way, as does the silver halide in photography except that instead of forming an image, the injected electrons produce electricity. Because of the minute thickness of the dye, each layer may not absorb very much light, but, like the leaves of a tree or the stacked thylakoid membrane found in photosynthesis, when added together, the many interconnected particles of the porous membrane can absorb 90% of visible light. The electrons lost by the dye are quickly replaced by the iodide in the electrolyte solution to produce iodine or tri-iodide, I_3^-, which in turn obtains an electron at the counter electrode after it has flowed through the load. The TiO_2 serves as the electron acceptor, and the iodide serves as the electron donor, with the dye functioning as a photochemical "pump." Several international companies are attempting to develop this technology and approach for practical photovoltaic modules, and they are expected to be available at a lower cost than Si-based technologies.

Several equations are used to describe the operation of the nanocrystalline DSSC. The incident photocurrent efficiency (IPCE) is given by

$$\text{IPCE} = a(\lambda) \bullet \phi_{inj} \bullet \eta_c , \tag{4.1}$$

where ϕ_{inj} is the injection efficiency. The other terms in this equation were defined in Chapters 2 and 3. The collection efficiency is, in part, determined by the rate constant for the "back" reaction, k_b, shown in Fig. 4.3. The units for k_b are s^{-1}. The injection efficiency is determined from the injection rate constant,

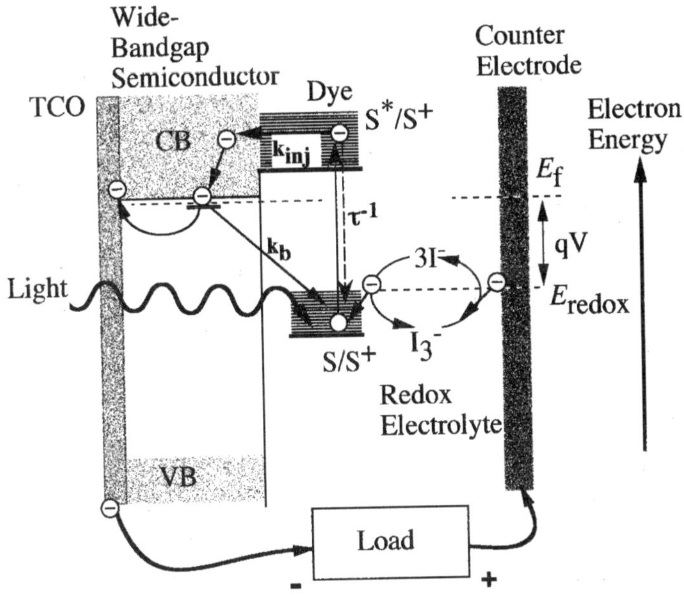

Figure 4.3 Energy band diagram for the dye-sensitized solar cell (DSSC). TCO is the transparent conductive oxide through which light enters the cell. The IPCE of a DSSC is determined by the injection rate constant, k_{inj}, and the back reaction rate constant, k_b, as well as the lifetime of the excited state, τ.

k_{inj}, and the excited state lifetime of the dye in the absence of electron injection, τ. Typical values for k_{inj} and τ are 10^{10}–10^{12} s^{-1}, and 50–100 ns, respectively. The injection efficiency is given by

$$\phi_{inj} = \frac{k_{inj}}{(k_{inj} + \tau^{-1})}. \tag{4.2}$$

This has resulted in IPCE values for Ru-based dyes that are 80–90% in the wavelength range from 400–750 nm (see Fig. 4.4). The I-V curves for the DSSC are described by equations of the form of Eqs. [3.1(a)] through [3.3(a)]. Typical values for these types of DSSC are V_{OC} = 0.8 V, J_{SC} = 16 mA/cm^2, and η = 9%. At present, the sealing, stability, and longevity of the nanocrystalline DSSC are being researched with the goal of practical applications.

With different solar cell technologies being developed as time goes on, how can an optical engineer view them in a common framework so that assessments and judgements can be made? The thermodynamic and optical framework described in the next chapter can be helpful toward creating a baseline by which all solar cells can be understood.

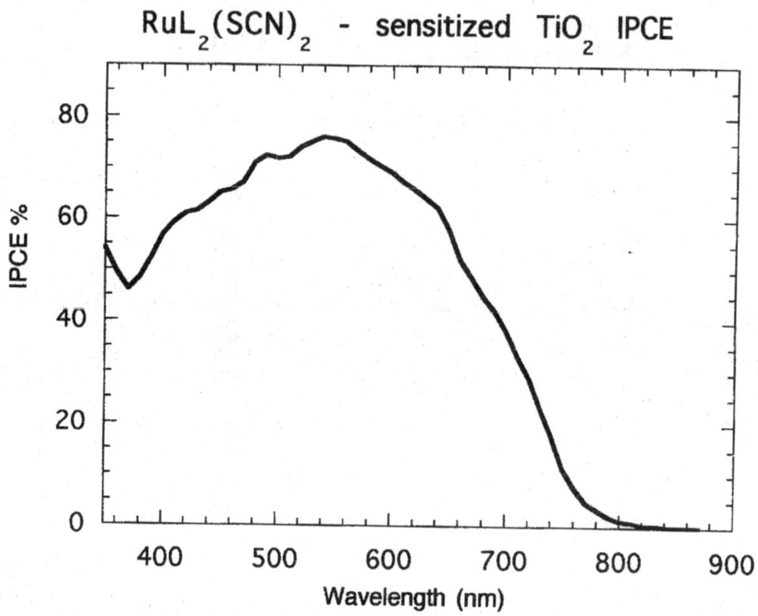

Figure 4.4 The incident photocurrent efficiency (quantum efficiency) for a typical dye-sensitized solar cell. The active area of the device was 0.3 cm^2 and the electrolyte used was I_3^-/I^- in acetronitrile. Since the transmission coefficient of the glass used for this DCCS was 0.8 in the wavelength range of 400–800 nm, this device exhibits nearly ideal injection and collection efficiencies. The integral of the product of this graph and the photon-flux density of Fig. 2.2(b) yields a current of 14 mA/cm^2. Under AM1.5 illumination, this device indeed produced J_{SC} = 14 mA/cm^2, together with V_{OC} = 0.8 V and FF = 0.7.

5

Generalized Model of a Solar Converter

5.1 General Thermodynamics of Light Conversion

Each type of quantum solar energy conversion system employs a conversion mechanism that can be radically different than the next. In each, photo-product and energy production can be varied and diverse, yet one optical concept may connect all quantum conversion systems—the time reversal of excitation by light is the production of luminescence. This luminescence characterizes the maximum work extraction and energy production in the converter. For any quantum system described, the absorber functions as a "pump" to increase the chemical potential on one side of a barrier. Recombination and "back" reactions (which include luminescence) compete with the work production in an external load. The general quantum-converter scheme is shown in Fig. 5.1. The amount of work done is given by the difference between the chemical potentials of electrons and holes on the acceptor and pump sides of the converter. This chemical potential difference is maintained by the light-driven pump due to the ability of the pump to maintain a difference between the concentration of excited electrons and holes on the acceptor and pump sides of the barrier. The barrier at the interface between the two sides of the device inhibits the back reaction. The concentration of excited electrons on the N-type (electron acceptor) side is n*, and the concentration of excited holes on the P-type side is p*. The equilibrium dark concentrations are given by n_0 and p_0. The chemical potentials on the two sides of the cell are given by

$$\mu_n = \mu_n - kT_0 \ln \frac{n^*}{n_0}, \qquad [5.1(a)]$$

$$\mu_p = \mu_p - kT_0 \ln \frac{p^*}{p_0}, \qquad [5.1(b)]$$

where k is Boltzmann's constant.

At a given temperature, T_0, the rate of work extraction is then given by the product of the difference in chemical potentials and the photo-current density, J, or

$$\frac{\text{Power}}{\text{Area}} = J\frac{(\mu_n - \mu_p)}{q} = J\frac{kT_0}{q}\ln\frac{n^*p^*}{n_0p_0} \quad . \tag{5.2}$$

In this view, solar conversion is treated like a chemical reaction involving the change in chemical potential of the charge carriers due to the incident photons.

For a time-symmetric system in which light is absorbed, one can also demonstrate that a luminescence efficiency of unity means that entropy production is minimized for the quantum solar converter, and that the power production (or rate of work extraction) will be maximized. Using the concepts of irreversible and finite time thermodynamics, Andresen and Berry [17, 18] determined an equation for the maximum rate of work extraction in any power-converting system. If we recall that the power output of a solar converter is equal to the time derivative of the work output (P = dW/dt), we can see that it is given

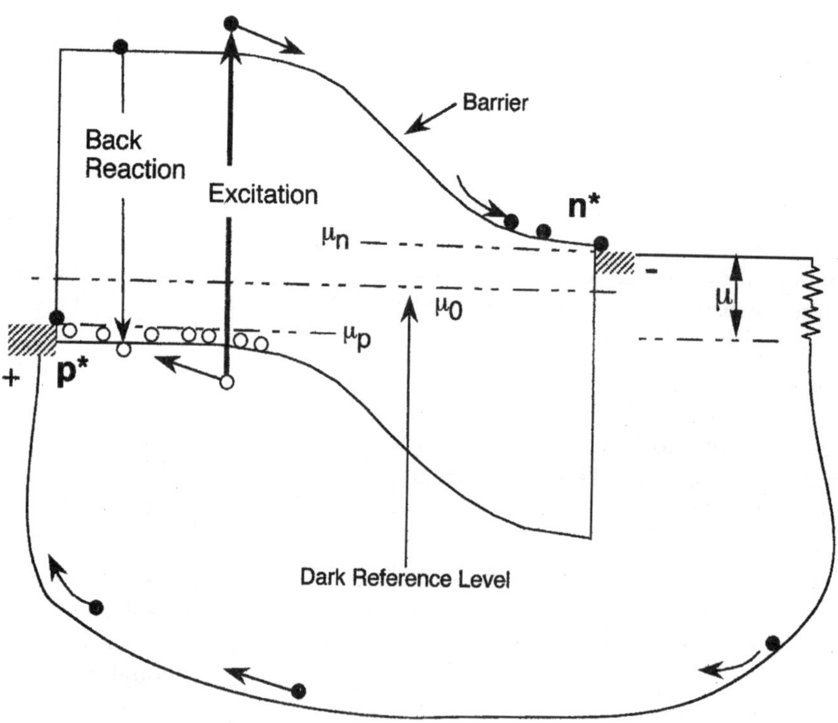

Fig. 5.1 A generalized quantum solar converter showing excitation by the light absorber. This "pumps" electrons to higher levels and allows a difference to develop between the concentration of excited states on the two sides of a "barrier." Work can be produced when these excited states return to the pump side through the load. Luminescence and "back reaction" compete with work extraction. Differences in chemical potential are indicated and are referenced to the level μ_0.

by the reversible work, W_{rev}, and the total entropy generation, S_{tot}. The entropy generation, when multiplied by the ambient temperature, defines the dissipation of available work. Starting with the Generalized Planck equation, and using the entropy involved in the absorption and emission of light, Parrot [14] derived an equation for a quantum solar converter of a similar form to that of Andresen and Berry. Combining the results from these authors, one obtains

$$\text{Power output density} = J \cdot \left(\mu_n - \mu_p\right)/q$$

$$\text{Power output} = \dot{W}_{rev} - T_0\dot{S}_{tot} = P_{abs}\left(1 - \frac{T_0}{T_S}\right) - \text{Dissipation}, \tag{5.3}$$

where the net absorbed power in the photoconverter is P_{abs}, and the dots over the symbols on the left-hand side of the equation represent time derivatives or rates. The ambient temperature is T_0 and the source temperature is T_S (i.e., the sun). These equations illustrate that a quantum solar converter maintains a state of nonequilibrium from which work can be extracted. It is this fact that allows photosynthesis and sunlight to maintain Earth (and life on it) in a state of nonequilibrium.

Since it can be connected to the chemical potential in the converter, the luminescent emission from a solar converter may be used to probe the thermodynamic state of the system. Outlined in the next section is the optical framework, which includes the Generalized Planck equation that can be used to understand the connection between chemical potential, luminescence, and current-voltage output in a quantum solar converter such as a solar cell. This is a basic theoretical framework that an optical engineer can appreciate and utilize.

5.2 Detailed Balance: The Generalized Planck Equation

In 1901, Max Planck derived an equation that extended the observations of Rayleigh and the calculations of Wien on the light emission from blackbody cavities. The equation was one of the first in the field of quantum mechanics. By using a model comprising a system of oscillators quantized in energy, together with thermodynamic principles, Planck was able to describe the full spectral output from a blackbody cavity held at a temperature, T_S. Although the Planck equation may be applied to thermal sources of light such as the sun or a light-bulb filament, recently it was realized that it could describe so-called "cold" emission from fluorescence or LEDs, which occurs at room temperature. The application of these models to a quantum photoconverter is possible due to four underlying assumptions:

(1) A thermalized two-level (or band) system may describe the upper limit to conversion efficiency.
(2) Bodies that absorb light must also emit light described by the Generalized Planck equation.
(3) Fluorescent emission is characterized by a nonzero chemical potential, μ.
(4) The measured product of charge and voltage in solar cells is limited to this chemical potential; therefore, the efficiencies predicted by this analysis represent the upper limits to solar conversion efficiencies in practical devices. This gives engineers a tool to predict device output given basic optical measurements.

Detailed balance calculations have long been used to obtain the ultimate efficiencies of solid-state solar cells. Originally pioneered by Shockley and Queisser, and extended to photochemical systems by Ross, these techniques consider the balance between the absorbed photon flux and the radiative and nonradiative recombination in the device. In the absence of nonradiative losses, the maximum efficiency for a single bandgap photovoltaic or photochemical conversion device, which accepts light from all angles, has been calculated to be 33% for the AM1.5 standard solar spectrum. It is worth noting that the maximum efficiency limit occurs within the range of bandgaps from 1.0–1.5 eV. In this way, materials in which the transition from low to high optical absorption occurs, between 1200 nm and 800 nm, are expected to produce devices with conversion efficiencies less than 33%. This analysis can also be applied to concentrated sunlight or other standard spectra with comparable results. The models used to calculate these limits take into consideration the optical absorption and solar-flux density, as well as the expected chemical potentials and voltages. These estimates have been applied to silicon solar cells and to photosynthesis from plants. For our purposes, the terms "chemical potential" and "voltage" can be loosely used interchangeably, although strictly speaking, the chemical potential represents the upper limit for the maximum product qV.

If the emitted light from a solar converter is strictly blackbody, work can be extracted *indirectly* by using a Carnot engine and by allowing the temperature of the converter to increase to above the ambient temperature. For a purely thermal energy converter, the chemical potential of the absorber does not change, but work can be extracted from the temperature difference. If direct conversion is required, as in photovoltaics or photochemistry, the free energy, μ, of the absorber must change. Figure 5.2 shows one thought experiment used to derive the Generalized Planck formula from Planck's original equation. Consider a system in which a photoconverter (shown at the right) is coupled through a selective bandpass filter to a blackbody emitter at temperature T_S. Both the converter and the source are housed in a perfectly reflecting (spherical) cavity, so the only exchange of radiation that takes place is through the filter. This filter is

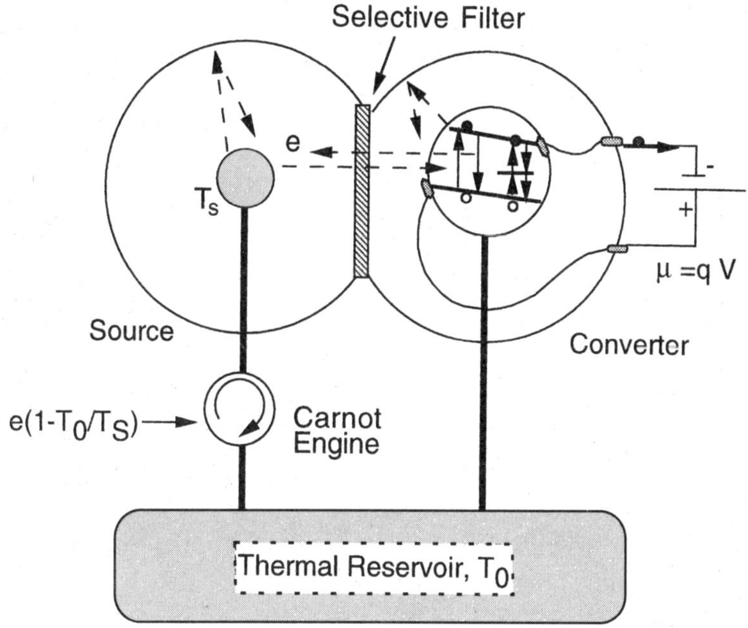

Fig. 5.2 Thought experiment used to determine the relationship between luminescent emission and photoconverter voltage. At steady state, the energy delivered to the blackbody source (at left), via the Carnot engine, to maintain the source at a constant temperature, T_s, is equal to the chemical potential, μ, generated by the converter (at right). Excitation and recombination processes are shown for the converter for single- and multiple-step processes.

perfectly reflecting for all photons except those in a narrow band centered on energy, e.

Photons emitted in this band by the source, or by the converter, pass through the filter without losses. The photons may be absorbed by the photoconverter that is connected to a thermal reservoir of temperature T_0. The absorption of light produces excited electrons in the upper energy level, which can decay, producing luminescent light that may be collected to produce work, fall into an intermediate trap, or into defect energy levels. This is shown schematically in the energy band diagram on the right-hand side of the figure. Each process occurs without the loss of energy from the system so that equilibrium is established between the various components of the system. Therefore, nonradiative recombination in the converter is neglected; this will be added later. The electrical work produced charges an ideal battery, or capacitor, at a voltage of $V = \mu/q$, where q is the elemental charge. In order for the temperature to remain at T_S, the energy loss from the source requires replacement via a heat pump connected to the thermal reservoir. This reversible Carnot engine delivers the necessary heat to the source, but requires an energy input of $e(1-T_0/T_S)$. In addition, the battery provides the energy for the converter via the (shaded) electrical contacts, producing a photon that travels back to the source. When zero net electrical current flows, equilibrium is established and an equal number of photons travel in both directions through the filter. From conservation of energy, we see that at a steady state

$$\mu = e\left(1 - \frac{T_0}{T_S}\right), \qquad [5.4(a)]$$

and

$$\frac{(e-\mu)}{kT_0} = \frac{e}{kT_S}. \qquad [5.4(b)]$$

Using this relationship in the Planck equation, and realizing that the radiance emitted by the source is equal to that of the converter, one obtains the so-called Generalized Planck equation

$$L_x(e,\mu,T_0) = \varepsilon(e)\frac{2\,n^2}{h^3\,c^2} \bullet \frac{e^3}{\exp\frac{(e-\mu_x)}{kT_0} - 1}, \qquad [5.5(a)]$$

where L is the spectral radiance (power per unit area per projected solid angle per photon energy interval). Other names for L include brightness and luminance. The constant, n, is the index of refraction of the medium in which the solid angle is measured. For example, for a silicon absorber immersed in glass or acrylic, the value of n is between 1.45 and 1.6. The constants h and c are Planck's constant and the speed of light in a vacuum, respectively. The constant k is Boltzmann's constant. When the photons from the total solid angle (180 deg) are included, a factor of π is added to the numerator of the above equation. Using convenient dimensions for the constants, the quantity $2\pi/h^3c^2$ is equal to $9.883 \times 10^{26}\ eV^{-3}\ s^{-1}\ m^{-2}$.

The chemical potential is given by μ, and represents the maximum free energy (in a thermodynamic sense) that may be extracted. The chemical potential divided by the charge per charge carrier, q, yields the maximum expected voltage from a solar cell. Since the photoconverter can be nonideal in its emission properties, an emissivity, $1 > \varepsilon(e) > 0$, is used in the Planck equation. The subscript "x" refers to the light source. In this text, "0" will refer to ambient radiation, the subscript "0R" will refer to the radiative recombination, and "S" will refer to solar or incident radiation on the absorber. The subscript "SC" will refer to the short-circuit condition. The Generalized Planck equation has the form one would expect for Boson gas, and photons are Bosons. The Wien approximation means the –1 in the denominator of the Planck equation is neglected. Using this approximation,

$$L_x = L_0 \exp(\mu/kT_0). \qquad [5.5(b)]$$

The Generalized Planck equation can be used for both the incoming and outgoing, or luminescent, light. A purely thermal radiator, such as the sun, can be viewed as a type of luminescent body with zero chemical potential and a high temperature. For example, the above equation can be used to describe the light from the sun if one sets the temperature to $T_s = 5762$ K. For solar radiation at Earth's surface, one multiplies this equation by f, which is the dilution factor ($0 < f < 1$) that represents the inverse square drop of the intensity from the sun's luminous surface to Earth. For 1000 W/m^2, f is approximately 1.6×10^{-5}.

Another way of looking at the Planck equation is to consider the entropy balance during the absorption of a photon. The loss of a photon from a beam of light is accompanied by a loss in entropy of e/T_S. If only μ is converted to work at ambient temperature T_0, then $(e - \mu)/T_0$ is the entropy transferred to the surroundings. Equating these two entropy values yields the Generalized Planck equation from the Planck blackbody equation. Alternatively, one can regard solar photons as having a chemical potential $\mu = e(1-T_0/T_S)$ equal to the maximum work, or free energy, they can produce at ambient temperature.

The emissivity, $0 < \varepsilon < 1$, measures how close the material comes to the ideal case. This will now be shown as a function of the material properties and absorber thickness. A quantity directly related to the emissivity is the absorptivity of the light. The quantum absorptivity, a(e), is the fraction of the incoming light that is absorbed by the material to produce an excited state. From a consideration of the detailed balance between the absorption and emission of light, it can be shown that the quantum absorptivity and emissivity are equal, just as are the thermal absorptivity and emissivity as expressed by Kirchhoff's law. Neglecting free-carrier absorption, the thermal and quantum absorptivities are approximately equal. Thus, a simple optical absorption measurement can be used to determine the emissivity. In order to prove this, we can define the "ideal" photon flux at energy e by

$$L'_x(e, \mu, T_0) = \frac{L_x(e, \mu, T_0)}{e \cdot \varepsilon(e)} \,. \qquad (5.6)$$

Dimensionally, L' is given, for example, in units of photons/sec/m^2. One can then describe the output light, L'_{out}, from an absorber as the sum of the emitted luminescent light and the nonabsorbed and reflected incident light for the geometry indicated in Fig. 5.3. This gives

$$L'_{out} = \varepsilon(e)L'_{0R} + L'_{in}\left[1 - a(e)\right], \qquad (5.7)$$

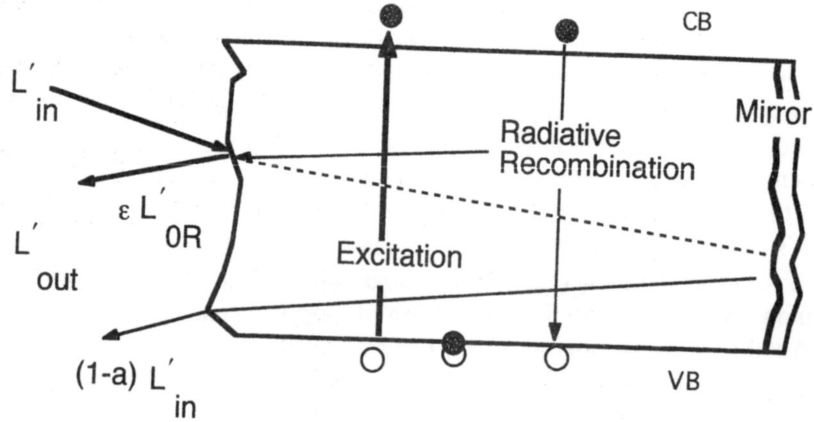

Fig. 5.3 A diagram used to determine the equality between absorptivity and emissivity.

if the incoming light is equal to the ambient blackbody radiation, $L'_{in} = L'_{0R}$, and the chemical potential in the Planck equation is zero ($L'_{0R} = L'_0$). The outgoing radiation must then balance the incoming, even in the presence of nonradiative losses. This means L'_{in} is equal to L'_{out}, which, in this case, gives us

$$a(e) = \varepsilon(e). \tag{5.8}$$

This equality is true since the absorption and emission are time reversible on a microscopic scale. Although we chose to evaluate the relationship between a(e) and ε(e) at ambient equilibrium, these quantities are characteristic parameters of the absorber at a given temperature. If the incoming flux is not equal to the ambient radiation, then the chemical potential is not zero, and from that the radiative photon flux, L'_{0R} , is approximated by its ambient value, L'_0, multiplied by a Boltzmann factor of exp μ/kT_0. We have shown above that the radiative losses are given by $a(e)L'_0$ under ambient radiation input, so that these losses are $a(e)L'_0\exp(\mu/kT_0)$ under solar illumination. Note that even though the probabilities given in Eq. (5.8) are equal, the photon fluxes for the output or input, as given in the Planck equation, will not always be equal. The input radiation is arbitrary, and the output luminescence is determined by the chemical potential. Figure 5.4 shows the luminescence predicted from the Generalized Planck equation together with measured values.

It should be emphasized that the use of the Generalized Planck equation is restricted to equilibrium radiation [19, 20, 22]. The predicted chemical potentials therefore represent the maximum values attainable in real systems. This is true since the equilibrium chemical potentials are always larger than those for nonequilibrium or irreversible cases. In these cases, the chemical potentials may be functions of time, position, and illumination, and may be coupled (through

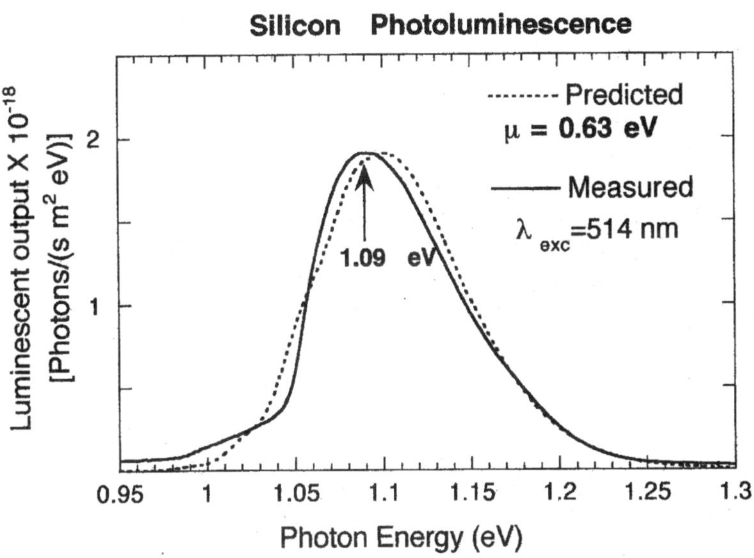

Fig. 5.4. Photoluminescence of a Si wafer used in solar cell manufacturing. Predicted (dashed line) and measured (solid line) photoluminescence spectra for silicon. The chemical potential used for the prediction is 0.63 eV. The prediction was made using the absorptivity calculated from the absorption coefficient data for Si and the simple optical model based on a polished wafer with a planar geometry.

feedback loops) to quantities such as the emissivity or temperature. In this analysis, simplifying assumptions are used in order to understand the fundamentals regarding maximum, or upper limits, for solar-conversion efficiencies based on optical considerations. In the next section, we relate the chemical potential to the incoming flux and photoluminescence efficiency.

5.3 The Luminescent Output

One may now connect the total luminescent flux density emitted by the photoconverter at ambient temperature to the chemical potential of the excited state. The photon-flux density is the number of photons per unit area per unit time. In this text, the various fluxes will be given the symbol Γ, with the subscript indicating the origin or nature of the flow. We can define the number of luminescent photons being emitted from a solar cell at ambient temperature as

$$\Gamma_{0R} = \int_0^\infty \varepsilon(e)\pi L'_{0R}\, de. \tag{5.9}$$

Note that a factor of π included in each integral is due to the integration over the full solid angle. Likewise, the ambient flux being absorbed, and also emitted, by the solar cell material is given by

$$\Gamma_0 = \int_0^\infty \varepsilon(e)\pi L_0' \, de. \tag{5.10}$$

The parameter Γ_0 is the radiated blackbody flux in the absence of an external light source.

Using the Wien approximation in the Planck equation and integrating over energy and solid angle, the total emitted luminescent flux density is approximated by

$$\Gamma_{OR} \cong \Gamma_0 \exp(\mu/kT_0). \tag{5.11}$$

The Wien approximation can also be used to obtain an approximate expression for the ambient photon current, Γ_0. Assuming a step-function absorptivity, the integration indicated in Eq. (5.10) yields

$$\Gamma_0 = \frac{2\pi n^2}{h^3 c^2}(kT_0)^3\left[\exp(-x_g)\right]\left(x_g^2 + 2x_g + 2\right), \tag{5.12}$$

where $x_g = e_g/kT_0$. A similar expression is obtained for solar input if kT_S (0.4966 eV) replaces kT_0 and a dilution factor, f, is utilized. This expression applies if the absorptivity is considered to be a step function at the bandgap energy. In other words, $a(e) = 1$ for $e > e_g$. In general, and especially for thin absorbers, the absorptivity is a function of the photon energy, and the integrals in Eqs. (5.9) through (5.11) must be carried out numerically.

5.4 The Relationship between Voltage and Luminescence Efficiency

From the Generalized Planck equation, it is possible to obtain the photoluminescent efficiency for a light emitter. For a slab of absorber material unconnected to a load, this luminescence efficiency, Φ, is simply the excess number of photons emitted from the material divided by the incident photons absorbed. This involves integration over energy and angle. This yields

$$\Phi_{OC}(\mu, T_0) \equiv \Phi = \frac{\text{Radiative Recombination}}{\text{Total Recombination}}. \tag{5.13}$$

The subscript "OC" refers to the open-circuit condition. For illumination from the sun, a photon flux Γ_S is absorbed. This yields

$$\Phi = \frac{\Gamma_{OR} - \Gamma_0}{\Gamma_S} \cong \frac{\Gamma_0 \exp(\mu/kT_0) - \Gamma_0}{\Gamma_S}. \tag{5.14}$$

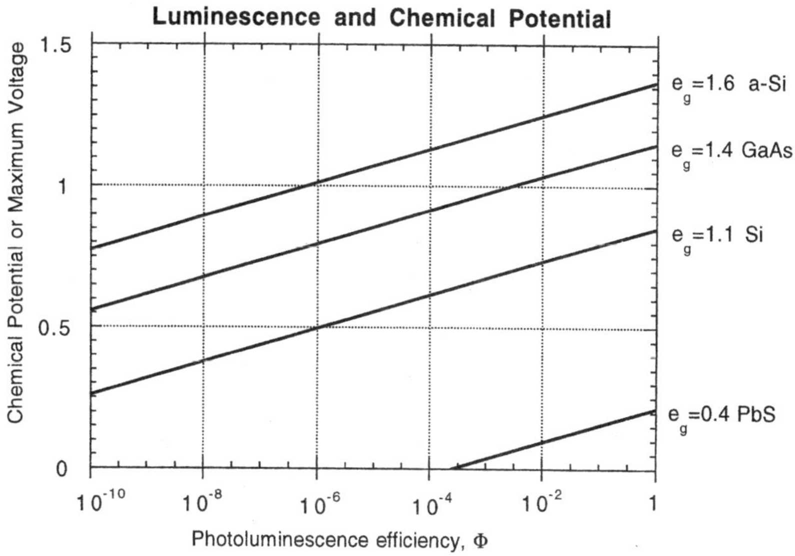

Fig. 5.5 Calculated chemical potential (charge times voltage) as a function of photoluminescence efficiency for various semiconductor materials. The calculation was made assuming solar illumination (AM1.5) and a step function absorptivity at the indicated bandgap energies.

The second term in the numerator in the above equation is the small background blackbody flux at ambient temperature. This yields the net recombination due to the illumination and excess charge-carrier density. Figure 5.5 shows the results of the above equations for various bandgap energies. For convenience, neglecting atmospheric absorption, the radiance of the sun, and ambient radiation, Γ_S and Γ_0, were evaluated as diluted blackbody radiation by setting $\mu = 0$ and $T_S = 5762$ K, and $T_0 = 300$ K, respectively [using Eq. (5.12)]. As can be seen, a logarithmic dependency of μ on Φ is found. A photoluminescence value of unity represents 100% conversion of the incident light into re-emitted light (at open circuit). Figure 5.4 shows the predicted and measured photoluminescence for a silicon wafer. One obtains from this plot and Fig. 5.5 that the photoluminescence efficiency for this Si wafer was 10^{-3}–10^{-4}. The analysis given above is a purely optical way of viewing a solar cell.

5.5 Current-Voltage Characteristics and Luminescence

To further emphasize the relationship between the maximum open-circuit voltage and photoluminescence efficiency we solve the approximation given in Eq. (5.14) for the chemical potential. This results in

$$\text{maximum chemical potential} \equiv \mu_{max} \approx kT_0 \ln\frac{\Gamma_S}{\Gamma_0} + kT_0 \ln \Phi, \qquad (5.15)$$

where Γ_S is the absorbed solar photon flux. This is similar to the result originally obtained by Shockley and Queisser, except for the term containing Φ. This equation contains the form of the diode equation used in semiconductor physics. It gives a correction for nonunity photoluminescence efficiency. It is seen from Eq. (5.15) and Fig. 5.4 that a small photoluminescence efficiency, on the order of 10^{-3} to 10^{-4}, results in voltage losses of 0.2 at room temperature. Similar to the previous analysis, one can obtain the full current-voltage characteristic for the solar cell in this way.

The derivation above can be expanded to include the effects of a current output and charge collection. This is possible by considering the detailed balance of electrons in the excited state shown in Figs. 1.3, 1.4, and 3.2. Series-resistive losses will be neglected here. Referring to the circuit diagram in Fig. 3.2, we see that current extraction, I, competes with radiative and nonradiative losses. The Planck equation allows for the determination of radiative losses, and their relationship to the generated chemical potential, μ. The nature of nonradiative losses is unknown. It is, however, instructive to assume that the nonradiative losses are of the same form as Eq. (5.11). This means that they depend exponentially on the chemical potential or voltage. This may seem ad hoc at first, but it is the same assumption used in standard semiconductor theory (Chapter 3). Using this assumption, we may represent the ratio of the nonradiative recombination flux to radiative flux, Γ_{0R}, as K, to obtain

$$\mathrm{K} = \text{Nonradiative losses/Radiative losses.} \tag{5.16}$$

When defining photoluminescence efficiency, from Eq. (5.13), note that

$$1 + \mathrm{K} \approx \frac{1}{\Phi}. \tag{5.17}$$

It should be emphasized that the parameter K, like Φ, is not a constant for the material (e.g., GaAs or Si), but is also dependent on the sample quality and defect concentration. In addition, the approximation in the previous equation is a result of the fact that the ratio of the nonradiative to radiative recombination may change as a current is extracted. However, for a given sample and extracted current, K can be considered a constant. This approximation allows one to balance electrons entering and leaving the excited state. Figures 1.3 and 1.4 illustrate the relationship between the various losses. The solar and ambient photon fluxes produce an output current after the losses from radiative and nonradiative recombination have reduced the flow. This yields the (detailed) balance

$$\begin{aligned}&\text{Ambient Incident} + \text{Solar Incident} = \\ &\text{Luminescent Photons} + \text{Phonons or Heat} + \text{Current extracted},\end{aligned} \qquad [5.18(a)]$$

$$(\mathrm{K}+1)\,\Gamma_0 + \Gamma_S = \Gamma_{0R} + \mathrm{K}\Gamma_{0R} + \Gamma, \qquad [5.18(b)]$$

where Γ is the current collected per unit area, for example in electrons/s/cm^2. The terms in this equation are defined by Eqs. [5.18(a)] and Eqs. (5.9) through (5.11). From Eq. (5.11), we have that the recombination flux is related to the chemical potential, μ. From this we obtain the equation

$$\Gamma = \Gamma_S - \frac{\Gamma_0}{\Phi}\left(\exp\frac{\mu}{kT_0} - 1\right), \tag{5.19}$$

which has the form of the well-known diode equation. If the electrical current, J, is desired, then the flux, Γ, is multiplied by the charge per particle, q. Likewise, the voltage is given by $\mu = qV$. Under open-circuit conditions, $\Gamma = 0$, and Eq. (5.19) yields Eqs. (5.14) and (5.15). Comparing Eq. (5.19) to the diode equation represented in Eq. [3.1(a)], $q\Gamma_0/\Phi$ is equal to the reverse saturation current for the ideal J-V curve. From this analysis, the $q\Gamma_0$ value for Si is on the order of 10^{-16} A/cm^2. The best reverse saturation currents for Si diodes are near 10^{-12} A/cm^2, so the Φ value for Si is on the order of 10^{-3} to 10^{-4}. Values in this range are indeed measured (Figs. 2.12 and 5.4).

The above equation can be understood as the balance between driving force, represented by μ, and kinetic or rate processes represented by the flux density, Γ, or the electrical current density, J. As the extracted current increases, the observed chemical potential will decrease and so will the luminescent flux. In the luminescence-based analysis presented by the equations of this chapter, the use of the absorptivity restricts one's knowledge to a gross overview of the particular sample. Therefore, the analysis of L, τ, and η_c in Chapter 3, and the luminescence-based type of analysis presented in this chapter are complementary and can be used together to understand the basics of solar cell device performance. Thermodynamic concepts such as chemical potential and kinetic concepts such as rate constants, lifetime, and collection probability may be used in parallel to build a more complete picture of a solar cell.

Using the type of analysis presented in this chapter, it has been calculated that the optimum bandgap for solar conversion is 1.1–1.4 eV and that PV conversion efficiencies of 33% are possible [19, 20]. For example, for an ideal c-Si solar cell, Eq. (5.15) and Fig. 5.5 predict an open circuit voltage limit of 0.85 V. Utilizing the AM1.5 data, shown in Fig. 2.2, one obtains a $q\Gamma_s$ value of 42.5 mA/cm^2 if one assumes a unity absorptivity and quantum efficiency for wavelengths of 360–1100 nm (compare this with Fig. 3.6). Using Eq. 5.19, one then predicts a FF of 0.7–0.8 and an upper limit for the c-Si solar conversion efficiency of $\eta = 27\%$. The best laboratory c-Si cells have conversion efficiencies of 21%, primarily due to the fact that even the purest silicon has a photoluminescence value, Φ, of 10^{-3}–10^{-4}. This means that experimental open circuit voltages for c-Si do not exceed 0.65–0.7 V.

An important relationship that can be derived from this type of analysis is the temperature dependence of the voltage. As the ambient temperature increases, so too will the radiative losses, Γ_{OR}. This will decrease the conversion efficiency. The derivative of Eq. (5.15) yields

$$d\mu/dT \approx -(1/T_0)(e_g - \mu) \tag{5.20}$$

if the step function absorptivity approximation, represented by Eq. (5.12), is used for Γ_{OR}. This derivative, which is equivalent to approximately −2 mV/K for silicon ($e_g = 1.1$ eV), is identical to that obtained using P-N junction theory. This lowering of the voltage with increasing temperature is observed experimentally for solar cells. This may be important for applications where a light concentrator is used to collect light onto the solar cell, since this may result in elevated temperatures. It should be pointed out that the optical-based analysis of solar cells presented in this chapter is a simplification, but it allows for useful insights into their nature and behavior, and it can be used as a guide for further technological developments.

6

Concentrators of Light

6.1 The Thermodynamic Limits of Light Concentrators

Almost everyone has had experience with a light concentrator. A hand lens focused on the ground on a sunny day demonstrates the ease with which sunlight can be collected and concentrated. What is not well known are the mechanisms for this concentration or the limits imposed by physics and thermodynamics. A survey of the literature over the past 20 years provides information on many different types of devices called concentrators. Some of these systems are as simple as the hand lens. Others, such as the fluorescent planar concentrator (FPC) seem, at first glance, to have radically different thermodynamic limitations. All of these systems, however different, are related in that they increase the number of photons on a surface or the irradiance (illumination) above the level occurring without the device. This is an advantage for solar-energy conversion and materials characterization since the receiver (e.g., a solar cell or thermal absorber) can be reduced in size relative to the total system. In this way, an area exposed to the sun can be covered by potentially cheaper, and technologically simpler, materials. High photon (flux) levels can also be used for the generation of high temperatures to produce steam, a photothermal reaction, or materials processing. In this chapter, the principles unifying geometrical optics and fluorescent concentrators are presented. General equations are developed and discussed in regard to their use in solar energy and solar concentrators.

A concentrator using only geometry, and not relying on a frequency shift, is called a geometrical, or passive, concentrator. A system that is concentrated only by a frequency shift is called a fluorescent, luminescent, or active system.

6.2 Geometrical Optics

6.2.1 General theory and sine brightness law

Consideration of the flux transfer in a geometrical system leads to an understanding of how a passive concentrator functions. A subscript 1 will refer to the entrance aperture, while a subscript 2 shall refer to the exit, or absorber, aperture. Considering a typical optical system (optical transformer) with entrance aperture A_1 and exit aperture A_2, light enters the system within a cone defined by $\pm\theta_1$ and leaves within $\pm\theta_2$ as measured from the optical axis (see Fig. 6.1). The

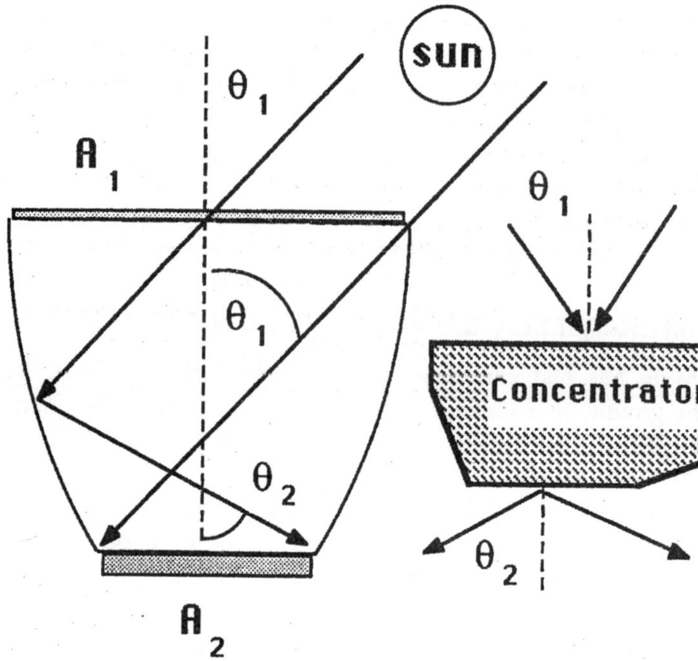

Fig. 6.1 A schematic of a light concentrator (right), and an example of an ideal concentrator represented by a compound parabolic concentrator (CPC). Light enters the optical system within a cone of rays defined by θ_1, and exits within θ_2. The light is concentrated because the area at the entrance aperture, A_1, is smaller than the area at the exit aperture, A_2.

brightness, luminance or radiance of the light, L′, is the flux per unit solid-state angle Ω, per unit-projected area. The flux incident on the top aperture from a Lambertian source such as the sun is then given by the integral of the radiance times the area and projected solid angle, or

$$\text{Input flux} = \int L_1' A_1 \cos\theta d\Omega = \int_0^\theta 2\pi L_1' A_1 \sin\theta \cos\theta d\theta = \pi L_1' A_1 \sin^2\theta_1. \tag{6.1}$$

A similar expression is obtained for the exit aperture with subscripts of 2. The concentration ratio, C, is given by the ratio of the flux density on the exit and entrance apertures, or

$$C = \frac{L_2' \sin^2\theta_2}{L_1' \sin^2\theta_1}. \tag{6.2}$$

In a geometrical system, concentration is obtained by conserving flux throughout the system, as well as the radiance of light ($L'_1 \le L'_2$). This means that as the beam area is decreased, the divergence, or angle, is increased to compensate. It can then be seen that the area is exchanged for the angle to achieve concentration. The maximum geometric concentration ratio is then given by

$$C \leq \frac{1}{\sin^2\theta_1} = \frac{\sin^2\theta_2}{\sin^2\theta_1}, \tag{6.3}$$

where the output angle is usually taken as 90 deg. The last two equations are the familiar sine brightness equations for ideal geometric flux transfer, which can also be derived using the phase space, or étendue, of the light. If the concentrator is made of a medium of refractive index, n, and the exit plane is immersed in this medium as well, then it is necessary to modify the concentrator equation. The edge ray θ_1 will be refracted to θ_1' in the concentrator, where $\sin\theta_1 = n \sin\theta_1'$ from Snell's law. For a concentrator with the exit aperture immersed in the medium, with θ_2 unchanged or unrefracted, the concentration is characterized by

$$C \leq \frac{\sin^2\theta_2}{\sin^2\theta_1'} = \frac{n^2\sin^2\theta_2}{\sin^2\theta_1}. \tag{6.4}$$

For convenience, the concentration is defined by the maximum incident external angle θ_1 and the final exit angle θ_2. Comparing this to Eq. (6.4), one concludes that $n^2L'_1 \geq L'_2$ for a passive system. This means that upon crossing into a medium with higher index of refraction, the radiation is confined to a smaller solid angle, and thus will have a higher radiance. The ramification of this will be discussed in a later section. The equations in this tutorial are presented for a 3D, or circularly symmetric, concentrator. For a 2D concentrator, where light is reduced in one direction (e.g., as in a trough concentrator), the concentration is the square root of the 3D value.

6.2.2 Examples of ideal geometric concentrators

Most imaging systems, such as Fresnel lenses and parabolic reflectors, fall short of the ideal limit by a factor of 4 or more. This is partially due to the fact that an image of the sun is transferred to the exit, as well as the flux, and that the exit angle is less than 90 deg. One type of concentrator that can approach the limit (equality) is the compound parabolic concentrator (CPC), which is an example of an "ideal" nonimaging concentrator. Shown in Fig. 6.1, a CPC is a concentrator with a cross section of two parabolic sections each tilted at θ_1' so the focus of each section is in the bottom corner of the other. Light incident on the entrance aperture reaches the exit plane after 1 to 2 bounces on the reflective sidewalls. This concentrator resembles a cone (or trough in 2D); in fact, an approximation of the CPC is often taken as a straight wall cone. As an illustration

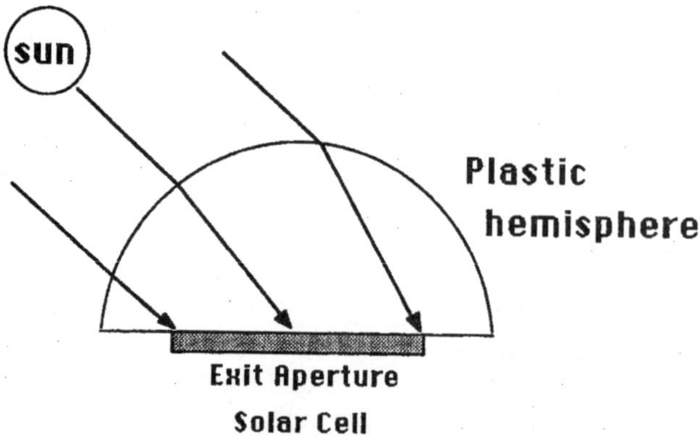

Fig. 6.2 A simple light concentrator that is useful for solar cells and detectors. A solar cell can be bonded to the bottom of a plastic (or glass) hemisphere such that diameter of the solar cell is W, and the diameter of the lens is nW. If the index of refraction, n, is 1.5, then concentrator will accept rays within ± 42 degrees and the concentration ratio will be approximately 2.2 ×. This means that the solar cell photocurrent (e.g., mA) and power output (e.g., W) will be 2.2 times that produced without the concentrator.

of the CPC, and of the use of these equations, consider a concentrator with $\theta_2 = 90$ deg, as is often the case, and $\theta_1 = 42$ deg. For the moment, let the refractive index, n, be unity (air). From Eq. (6.4), the concentration limit is 2.2 suns or 2.2 times the incident. If placed in a fixed orientation, the concentrator could collect for approximately 6 h during the day, with only minimal adjustment for season. If this concentrator is filled with an oil of refractive index 1.5, then a ray incident at 90 deg on the entrance aperture will be refracted to 42 deg. The concentrator now functions with $\theta_1 = 90$ deg, but the concentration ratio is unchanged at 2.2, which is n^2. This is because the initial choice of the input angle, 42 deg, is the critical angle, $\theta_c = \sin^{-1}(1/n)$. The concentration is over a 2π solid angle and the system will accept fully diffuse radiation (e.g., on a cloudy day).

Other systems have been devised that operate in this n^2 limit for concentration. One system is a hemispherical lens of diameter nW, where W is the width of the attached absorber or solar cell, and n is the refractive index of the dome (see Fig. 6.2). Light enters the dome and is refracted to the exit plane. This concentrator has a variable entrance aperture because different portions of the device are used at different incident angles. For any input angle, the rays that intercept the dome but would otherwise have crossed the diameter area of the hemisphere will reach the immersed exit aperture. This is one of a few imaging systems that operates close to ideal concentration limits. In fact, if one views the solar cell through the dome at any angle, one sees a 2×-magnified image. The hemisphere lens has no spherical aberration or coma, and is related to the aplanatic lens used in microscopy. The concentration is essentially a constant—near 2–2.2 suns, with respect to entrance angle. Although the concentration ratio is essentially constant, the output of a solar cell falls off dramatically with the incidence angle. If sunlight is used for the experiment, the output varies with the cosine of the angle, but the concentration ratio is unchanged. This is an important and general conclusion in concentrator technology—only radiation present at its entrance aperture is concentrated. Since the illumination at the entrance varies with the entrance angle, the output also varies. In practice, as the angle increases, the transmission losses must also be included.

Another example of a concentrator involves the PV itself. An n^2 concentrator is created when solar cells are attached to a transparent plate with a white scattering surface present on the underside in areas not bonded to solar cells (see Fig. 6.3). Light refracts into the plate and scatters off the bottom surface in all directions. A fraction, $\sin^2\theta$, of this light escapes, but some light undergoes total internal reflection and is collected at another solar cell bonded to the plate. This system is nonideal, but can achieve a practical concentration of 1.7 suns, and is used by some photovoltaic panel manufacturers.

Ideal geometric concentrators have been built that function in the high concentration range (10–100,000 suns). These systems are related to the simple CPC, but use reflective hyperbolic or straight sidewalls, and a lens at the entrance

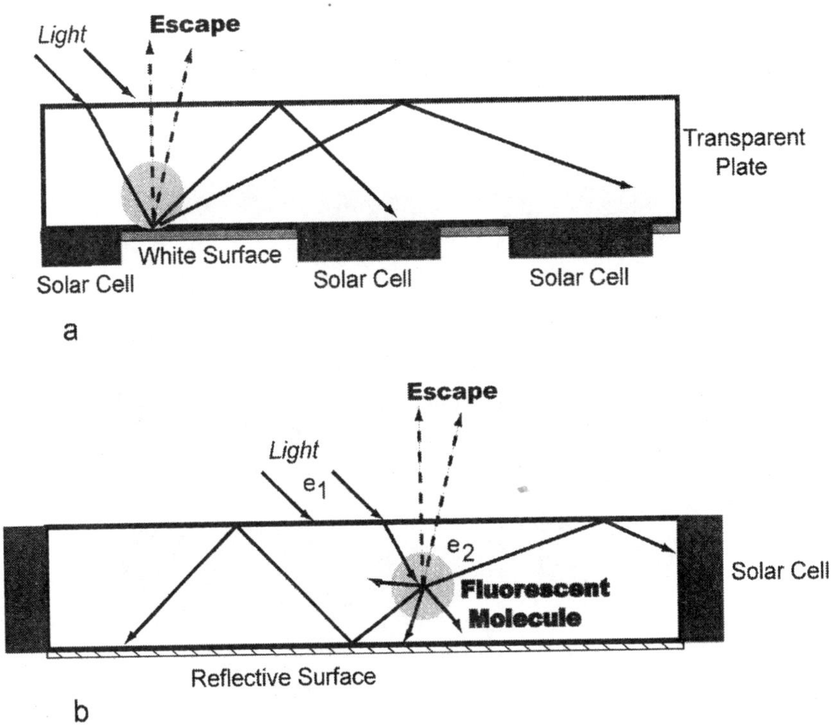

Fig. 6.3 Concentrators using flat plates, a) White transparent plate, b) plate with fluorescent dye molecules that absorb light at photon energy e_1 and emit it at a lower photon energy e_2. Light is shown undergoing total internal reflection at the top of the plate and reflection at the bottom. The light is collected by solar cells.

aperture in order to reduce the total system length. High-concentration-ratio systems require solar tracking, since the acceptance angle is small, but have achieved 84,000 suns and 7.2 kW/cm^2 irradiance in experimental facilities. In the case of high concentration, θ_1 is limited by the sun's angular size (0.27 deg–0.26 deg). For n = 1.5, this means that a concentration of 100,000 suns is theoretically possible using the above equations. These high-flux levels could be used for a solar-pumped laser, or in the destruction of hazardous chemical wastes. It is also worth noting that since the nonimaging concentrators operate near the equality found in Eq. (6.4), they can also be used in reverse to control the angular-output range of a light source. This property makes them interesting for use in collimating light sources such as light-emitting diodes.

Due to high irradiance levels, concentrator solar cells require an intricate current collection grid in order to avoid high series resistance losses (see Fig. 6.4). For "one sun" or low-concentration PV modules, the grid pattern instead looks more like that shown in Fig. 1.1. The grid pattern on most cells resembles the pattern of veins on a leaf. This should not be surprising, since both the solar cell and leaf are similarly limited by the need to collect photo products (e.g., sugars or electrons) across large surfaces.

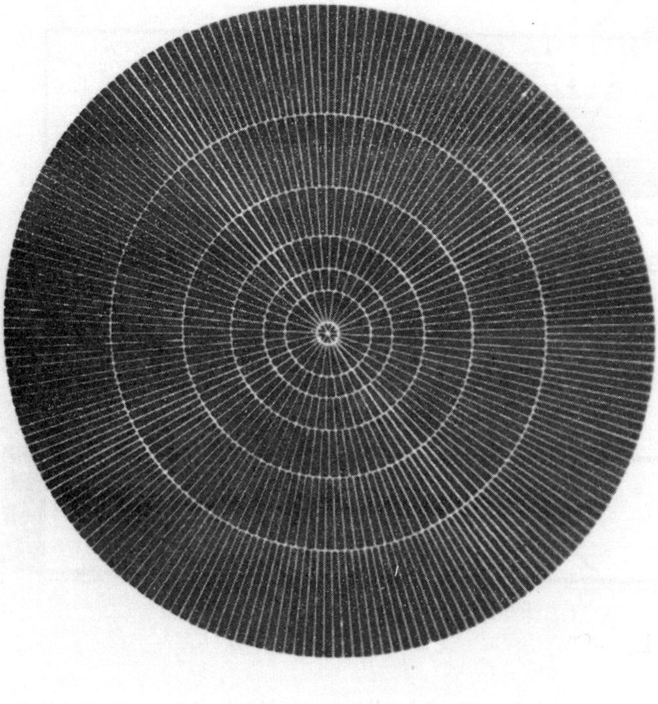

Fig. 6.4 A metal contact grid pattern for a concentrator solar cell (top contact).

6.3 Active Concentrators (Luminescent Systems)

6.3.1 Theory of luminescent systems

From the previous section, it would seem that the limit for diffuse radiation concentration is n^2. This is experimentally surpassed in the case of the fluorescent concentrator, which concentrates diffuse, as well as direct, radiation to levels above this. In the current technique, a plastic or dielectric material is doped with an organic dye or fluorescent inorganic molecule (see Figure 6.3). Light is absorbed at one energy and is Stokes-shifted or re-emitted to a lower energy (from the blue wavelength to the red). A portion of this light is trapped in the plate via total internal reflection, and it can be collected by the absorber plane or exit aperture. In this way, the system resembles the white painted transparent plate described in the previous section. Both the entrance and exit angles are 90 deg. Another similarity is that since the absorption of the incident light is not dependent on the angle, the system can operate at all incidence angles. What is different is that the process is physical or quantum, and does not rely on geometrical optics. This means that the concentration is not limited by the geometrical optics described in the previous section. In the next section, the radiance, L, will be considered in a small spectral band of energy, e, where e is the photon energy, $h\nu$. The process, which is really a chemical reaction occurring via the fluorescent molecule, is

$$\text{photon}(e_1) \leftrightarrow \text{photon}(e_2) + \text{heat}, \tag{6.5}$$

where $e_1 > e_2$. The fluorescent concentrator operates like an optical heat pump. The radiance at one energy is increased by changing some of the incoming energy to heat. To solve for the concentration ratio as a function of the Stokes shift, we consider the Generalized Planck equation [Eq. 5.5(a)]. Solving for the chemical potential for an index of refraction of 1.0, one obtains

$$\mu_x = e_x - kT_0 \ln\left(\frac{2}{h^3c^2}\frac{e_x^2}{L'_x} + 1\right), \tag{6.6}$$

where the x subscript indicates the process. By equating the chemical potentials for the absorbed and emitted photon, and solving for the flux, one can obtain the approximate concentration ratio valid for incident illumination below 100 suns. This equation is given by

$$C = \frac{L'_2}{L'_1} \approx \frac{e_2^2}{e_1^2} \exp\frac{e_1 - e_2}{kT_0}. \tag{6.7}$$

Theoretical concentration ratios for $e_1 = 2.14$ eV and $e_2 = 2.02$ eV are below 100 suns, but self absorption and light trapping losses limit actual concentration

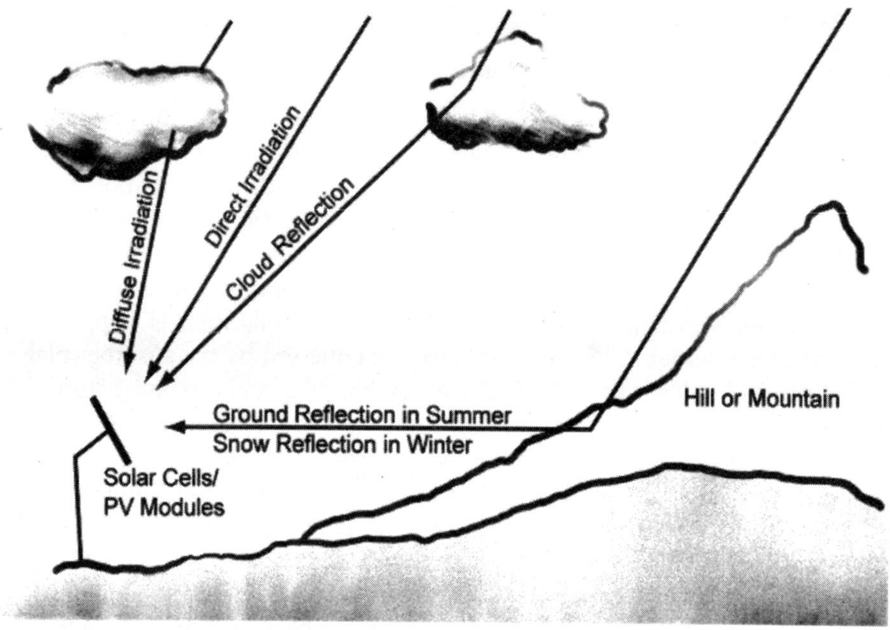

Fig. 6.5 Interception of light by a solar cell in the environment.

values to 10 suns. Still, the outdoor stability of some perylene dyes, an organic compound, warrants further study in this type of concentrator. The equations in this chapter describe the basic operation of man-made optical concentrators. Sometimes, the environment itself can provide the concentrator.

6.4 Light Enhancement in the Environment

Figure 6.5 shows how a solar cell or PV module would receive light in the environment in which it is mounted. As is shown in the diagram, both direct and diffuse light can be collected if the module is of the "flat plate" type or is a concentrator system of low C value. Careful selection of the orientation and mounting of the PV panels can often allow them to collect additional light beyond mere direct radiation from the sunlight. Sunlight reflected from clouds, the ground, or from snow can be collected. In this case, the environment itself can be a kind of light concentrator. This reflected light is due to the "albedo" of the ground or clouds. This term is used to describe the reflectivity of these natural surfaces. Because of this albedo effect, one must be careful not to underestimate the amount of light that is actually being received by the solar cell when it is tested outdoors. One may be fooled into thinking that the solar-power input is 1000 W/m^2, when it can actually be as much as twice this value for short periods. Accuracy in outdoor solar conversion efficiency measurements is best accomplished by using a pyranometer that actually measures the diffuse and direct input power (in W/m^2) at a given location. This information can be collected along with the power output to determine the system's performance.

PV panels can be mounted on racks on rooftops, on buildings as awnings or as structural units. The latter concept is called "building-integrated photovoltaics" and can lower the overall cost of the PV system since the modules serve a dual purpose of structural material and power generator. With photovoltaic costs of \$500/m^2, stone costs of \$800–2000/m^2, glass wall costs of \$560–800/m^2, and stainless steel wall and frame costs of \$280–400/m^2, it is not difficult to understand why some buildings worldwide are starting to use PV in place of some traditional building materials. We see that in mounting the PV panel, the concentration of light (optical gain) can also occur when reflected light from one module strikes another. Thus, the design of module mounting is another aspect of solar cells that can benefit from optical considerations. It should be noted that high-concentration solar systems with small acceptance angles have a limited ability to collect this additional light. Passive and geometrical concentration systems with high C values require mechanical solar tracking, but offer potential cost-saving advantages due to the fact that the solar cell area can be reduced relative to the area of the lens or mirror of the concentrator. There are several descriptions of such concentrator systems related to PV [9, 20, 21, 23].

7 Economics of Photovoltaic Cells and Systems

7.1 The Basics of PV Economics

The economics of PV devices are also related to the optics. The cost of photovoltaic materials is often expressed on a per-unit-area basis, but the modules are often sold based on cost per peak watt potentially generated. To convert the cost per square meter to the cost per peak watt, the following equation is employed:

$$\$/W_P = \frac{\$/m^2}{\eta \cdot 1000\ W_P/m^2}, \tag{7.1}$$

where η is the conversion efficiency. A 12% efficient module with a cost of $\$400/m^2$ yields a cost per peak watt of \$3.30. In a simplified economic analysis, it is desirable to estimate the return on the investment made for a particular material used as part of the photovoltaic system. The time for an investment return (payback time) on the PV module, of cost $\$/m^2$, is related to its efficiency, and the cost at which electricity is sold on the market, in \$/kWh. The payback time is expressed as follows:

$$\text{Payback time} = \frac{\text{Cost } \$/m^2}{\eta \cdot \dfrac{5\,\text{kWh}}{\text{day} \cdot m^2} \cdot \dfrac{365\,\text{day}}{\text{year}} \cdot \dfrac{\text{electricity}\,\$}{\text{kWh}}}. \tag{7.2}$$

For example, the payback time for a $\$75/m^2$ module of 10% efficiency at an electricity-selling price of \$0.08/kWh is approximately 5 years. Commercial crystalline-silicon (c-Si) solar cells are currently produced by growing single crystal ingots using very pure materials. The silicon is sliced and processed with dopants and methods that are very similar to silicon technology used for making integrated circuits and computer chips. This process is not conducive to high-volume production rates or large areas per unit time. The c-Si costs are currently $\$400–500/m^2$. Newer thin-film technologies are based on materials such as (noncrystalline) amorphous-silicon (a-Si), or on polycrystalline CdTe or $CuInSe_2$. These thin materials can be directly deposited on glass and interconnected using laser or photolithography patterning (see Fig. 1.6).

7.2 Estimated Solar Module Cost

To estimate the costs of a PV module, consider how much a hypothetical thin-film solar cell material costs (see Table 7.1). In the production of a PV module, the materials, and the fabrication must be considered. Unlike crystalline solar cells such as c-Si, major costs for thin-film technology include the glass and the manufacturing and operation costs.

Table 7.1 The cost of the hypothetical PV module.

	Costs
Solar cell materials and glass	\$50.00
Production overhead: equipment depreciation, indirect and other direct materials	\$5.00
Labor: direct and indirect (including assembly and testing)	\$1.00
Encapsulant or sealant	\$2.00
Frame and electrical interconnects	\$2.00
Additional protective cover or Tedlar backing	\$2.00
Profit, Interest due on loans	\$2.00
Total module cost	**\$64/m^2**

The direct costs, such as those for tools and labor, are related to the actual production of the module, while the indirect costs incurred for such things as accountants, rent, and computers are volume insensitive. This calculation assumes a 5–10 MW_p/year factory with 100 employees, and a capital cost of equipment of \$17,000,000, housed in a 2000 m^2 facility. The module costs are determined primarily by the cost of the conductive glass, and the production overhead. To estimate the cost per peak watt, one relates the cost per unit area with the power produced, which depends on the solar conversion efficiency and the peak solar illumination as described above. For a PV module at an efficiency of 8%, power would be produced at \$0.80/W_p if the module cost is \$64/m^2. For a 10% efficient cell, the cost would be approximately \$0.64/W_p. The estimated solar module production cost for a factory producing 10 MW_p per year of 10% efficient solar modules is found to be less than \$2/W_p if the module cost per square meter is less than \$100 As a comparison, the module costs for single crystal-silicon (c-Si) cells are now approximately \$5–8/W_p.

7.3 Economics of Photovoltaic Systems

To produce useful power in a commercial application, one must consider the average illumination, instead of the peak, as well as the additional costs of land, batteries, support structures, *and* the lifetime of the panel. If these "balance of systems" (BOS) costs are considered, the cost of power produced with thin-film solar cells would be \$0.08–0.11/kWh, assuming a 10% efficient module which lasts at least 15 years under the illumination found in the western United States. Note that this cost lies in the range of electricity for conventional fossil-fuel-based systems. Thus, the above analysis demonstrates that the thin-film solar cells, if proven stable over 15 years, could represent a viable solar energy option. There is more to a photovoltaic system than just the module. Balance of systems costs such as the mounting, wiring, and power conditioning must be considered, as well as the operating and maintenance costs. When these factors are taken into consideration, a rough cost per generated kilowatt hour can be estimated. The cost of electricity generated by solar cells can be estimated from the equation:

$$\frac{\text{Cost}}{\text{kWh}} = \frac{\left(\text{Cost of system } \$/\text{m}^2\right)\cdot \text{amortization}}{\text{kWh produced each year}} + \text{O\&M}. \tag{7.3a}$$

Putting in all the relevant terms, Eq. [7.3(a)] becomes

$$\frac{\$}{\text{kWh}} = \frac{\text{Module } \$/\text{m}^2 + \text{Mount } \$/\text{m}^2 + \left(\text{PC } \$/\text{kW}\right)\cdot \eta \cdot 1\,\text{kW/m}^2}{\eta \cdot \dfrac{5\,\text{kWh}}{\text{day}\cdot \text{m}^2}\cdot \dfrac{365\text{day}}{\text{year}}} \cdot \text{amortization}\left(1 + \text{Indirect}\right) + \text{O\&M}. \tag{7.3b}$$

One should note that if a concentrator is used, the module cost is divided by the concentration ratio, and the mount costs, O&M costs, and input kWh are adjusted accordingly. The per-unit-area cost of solar cells seems to be the largest and most variable cost item in a photovoltaic system. For a relative comparison, one can use the following numbers that are believed to be reasonable for near-term thin-film PV technologies:

(1) a module cost of \$75/m^2,
(2) a mounting cost of \$50–75/m^2 (mount, land, wiring),
(3) PC cost of \$170 per peak kW
(power conditioning, battery storage, and dc-ac inverter), and
(4) an indirect cost of 30% for architect and engineer fees, along with interest during construction.

The costs of the PV system are paid off over the lifetime of the project. The amortization rate is calculated from the real discount rate of i, and a PV lifetime, N:

$$\text{amortization} = \frac{i}{\left[1-(1+i)^{-N}\right]}. \tag{7.4}$$

The amortization rate is 0.07–0.1 for N = 15–30 years and a discount of 6–12%. For a solar insolation of 4.4–5 kWh/day/m^2 (1600–1800 kWh/year/ m^2), and an operating and maintenance (O&M) cost of $0.005/kWh, the electricity costs can thus be estimated. The cost of conventional electricity in U.S. dollars is $0.06–0.12/kWh. Table 7.2 shows the results of this simplified analysis and indicates that although solar cells of 15% efficiency that last for 15 years can be competitive with fossil fuels, cells of less than 8% efficiency with lifetimes of under 15 years will probably not be economical. To compare present costs and future needs for PV, see Table 7.3.

Table 7.2 Cost of a photovoltaic system.

Electricity cost as $/kWh				
PV efficiency, η	8%	10%	15%	20%
N=30 years	0.093	0.077	0.056	0.045
N=15 years	0.13	0.108	0.078	0.062

Table 7.3 Present and future PV system cost breakdown.

Aspect of PV system cost	**Now**	**Needed**
Module cost	$450–$500/m^2	$150–$75/m^2
Area-related BOS	$135/$m^2$	$150–$75/m^2
Power conditioning	$200/kW ($20/m^2)	$100/kW ($14/m^2)
Module efficiency	10–15 %	15 %
Cost of DC electricity	$0.20/kWh	$0.04/kWh
AC cost with storage	$0.30/kWh	$0.07/kWh
Module cost/W_p	$5–$7/W_p	$2–$3 /W_p

The analysis in Table 7.3 outlines the basic factors involved in the costs of both the PV cells and the PV systems and illustrates the areas for improvements. This analysis does not include the cost of capital in the form of loans, nor does it include the social and environmental benefits of solar energy converters compared to conventional power sources such as fossil fuels or nuclear power.

7.4 Economics of Solar Energy in the World Economy

Energy policy often neglects to consider the complex interplay between science, technology, and history. Once a set of technologies is chosen (e.g., fossil-fuel energy sources), it is difficult to move to more attractive, newer technologies without large investments and efforts. Scientists and engineers often do not see the economic and policy aspects of their new developments. Conflicts in the Middle East and elsewhere will continue to emphasize the disadvantages of reliance on nonrenewable energy resources. As concerns over energy resources and the consequences of pollution become more important internationally, the use of renewable energy will be increasingly important. Alternative energy technologies alone cannot ensure sustained economic development and growth. Likewise, policies that do not consider energy sources in their technical, historical, environmental, and economic contexts are doomed to fail in the end. Many questions exist as to whether renewable energy is up to the task of supplying a significant fraction of the world's energy needs. The question becomes how to promote renewable energy, and how to use both technology and policy to ensure that growth in energy supplies can meet future demand in a sustainable way. Discussions on renewable energy often focus on two questions: (1) can renewable energy supply our energy needs, and (2) can it be done economically? There is ample evidence that the answer to both of these questions is "yes."

On the exterior glass of a 48-story building located at 4 Times Square in New York City, 15 kW are being generated from sunlight using photovoltaics, otherwise known as solar electric panels. On the fourth floor, two 200-kW fuel-cell generators silently and efficiently provide energy for the entire building. The state-of-the-art building is also equipped with many other renewable energy and energy efficient technologies. In an article entitled "A realizable renewable energy future," by Dr. John Turner of the U.S. Dept. of Energy's National Renewable Energy Laboratory in Golden, Colorado, it is stated that, "PV technology has the ability alone to provide all of the energy needs of the United States." His calculation assumes a 10% solar-to-electrical-system efficiency, well within the efficiency capacity of today's technology and the use of fixed flat-plate collectors that are now the PV industry standard. Dr. Turner's conclusion has been asserted several times before but is nonetheless noteworthy. Using standard PV technology, a square 161 km (100 miles) on a side, in 1 year, would produce the energy equivalent to that used annually in the entire United States. Although 25,921 km^2 (10,000 square miles) is a large area, it is less than one quarter of the area that the U.S. has covered with roads and streets, and is much smaller than the area in the U.S. devoted to cropland. This area could be located in one place (e.g., a desert area), or distributed on every suitable roof or area. It should be noted that if wind power is added to a country's energy mix, the required area for PV is reduced. For example, the San Gorgonio pass in southern California has the wind energy equivalent of seven nuclear power plants. Wind power is now cost competitive with electricity generation from fossil fuels in several areas, and the installed wind-turbine capacity is expected to grow by more than 25% per year over the next few decades.

The cost of solar photovoltaics has dropped from over $100 per peak watt in the 1970s to under $6 per peak watt in 2002. Types of solar panels have diversified and now include 100–200 W modules of crystalline silicon, amorphous silicon, cadmium telluride, copper-indium diselenide, and others. The light energy to electrical-energy-conversion efficiency of these panels typically ranges from 10% to 15%, with steadily increasing values over the last 20 years. With continued interest and investment, the trend in increasing efficiency and decreasing costs is expected to continue. Applications of PV include satellite power, power for small consumer appliances, remote residential and industrial power, telecommunications, cathodic protection, water pumping and treatment, military and grid connected systems. Markets have grown at rates in excess of 15–20 % per year, and are expected to continue to grow at these rates or higher in the foreseeable future. The growth of energy markets in the developing world has prompted energy giants such as British Petroleum, Kyocera, Siemens, and Shell to purchase PV manufacturers. According to the PV industry analysts, more than 100 MW of solar panels are produced and shipped worldwide each year. Still, this puts the cost at several times the cost of conventional energy sources. Solar power and the other renewable energies are now cost competitive in many locations. Unlike conventional energy sources, these resources do not need to follow price spikes that have plagued the consumers of conventional energy sources. Once the initial cost of the installation is financed, the energy flows constantly and reliably. When the wind is not blowing or the sun is not shining, energy customers can buy energy from their local utilities. When the renewable resource (e.g., sun or wind) is available, however, excess energy can be put back into the grid for other customers to use. This storage concept is called "net metering" or "distributed generation," and is possible because of efficient dc-to-ac converters (electronic inverters) that allow the small residential renewable energy systems to be "in phase" with the ac electricity on the transmission lines. Many utility companies give their customers full credit for the energy they generate. This could turn every building into a mini-power plant. The point is clear: we can collect and convert more than enough renewable energy to power our society. So if technology alone is not limiting the use of renewable energy, then what is? Perhaps it is the economics or the way we view the economics.

One important economic issue that is often neglected when formulating energy policy is the subsidization of energy production, whereby government action is taken to influence energy market outcomes, whether through financial incentives, regulation, research and development, or public enterprise. Consideration of subsidies, in general, is becoming increasingly important for international trade, and the World Trade Organization (WTO) is discussing it. Many regulations placed on power companies and energy providers often act as subsidies. In some cases, those regulations and certain subsidies actually protect the environment, and the consumer. Subsides can favor certain energy industries and skew the economics so that one energy source looks unfavorable when it actually may not be. Norman Myers and Jennifer Kent [27] claim that today, subsidies worldwide are strongly weighted against nonpolluting renewable energy sources. Many studies assert that continued investment in renewable energy, such as solar energy, could stimulate economic growth worldwide [24-26]. Subsidies and political factors, however, often hide the full costs of conventional sources of energy. In the U.S., it is estimated that energy subsidies total \$32–36 billion, approximately 90% of which goes to fossil fuels and nuclear power [27]. Another example comes from the Dominican Republic, where grid-connected electricity costs to consumers are \$0.11/kWh, typical of many developed nations. This cost does not, however, reflect cost (e.g. \$11/kWh) to string the transmission wire needed to reach the rural villages, nor does it reflect the limited supply of government money needed to provide such grid connection service to only a few scattered villages. For coal alone, the most subsidized energy source, subsidies total somewhere between \$37 and \$51 billion worldwide, with \$17 billion from the former Soviet Union and \$6 billion in China and India [27].

Even in light of these subsidies, the prejudice often persists that solar electricity is expensive compared to that from fossil fuels and nonrenewable alternatives. This assumption is clearly not true in many developing nations. For example, a study conducted by researchers at Sandia National Labs concluded that in the Dominican Republic, PV is the logical choice for remote rural villages. People there use energy sources such as kerosene and candles for lighting, automotive batteries to run televisions, and dry cell batteries for radios and consumer electronics. Although the cost for the purchase of each of these energy sources may be low, the cost per unit of energy is, on average, as much as \$2.00/kWh. In contrast, a small village photovoltaic system can reliably and consistently supply the same loads at costs less than \$0.75/kWh, which is higher than grid connected electricity, but still lower than the villagers would pay otherwise. Unfortunately, most of the 2 billion or more people in the world who could benefit from solar PV technology cannot afford to pay the high up-front costs associated with it.

Although the fuel is free and the maintenance is minimal, one of the major limitations of solar energy technologies is the necessity for large capital investments to be made at the beginning of a project. This has discouraged its use in many small applications worldwide due to the lack of available initial funds. In addition, energy prices are often too low for solar (and other renewable) energy to compete on economic grounds, in part due to explicit or implicit subsidies for conventional energy. If risks are perceived as being too high, investments by local banks and institutions will not be made. Several organizations are meeting this challenge and raising awareness in lending practices. For example, the Solar Electric Light Fund (SELF) in Washington, D.C., has provided tens of thousands of low-cost loans to small PV projects in Vietnam, India, China, and other developing countries. The World Bank and the Global Environmental Facility (GEF) have encouraged many international renewable energy projects, including those using solar energy [26]. In 1992, the World Bank established the Asia

Alternative Energy program, which helps to bring renewable energy and energy efficiency practices into the forefront in lending programs in Asia. In China, a rural energy and greenhouse gas mitigation study conducted in 1994 led to the current Renewable Energy Development Project underway there. In India, a PV program for home and commercial use was started in 1992. In 1997, similar programs were started in Indonesia and Sri Lanka for PV home systems in rural off-grid applications. Argentina followed in 1998, with programs in Mexico and the Philippines starting in 1999 [26]. These programs are typically funded in the \$50–150 million range and provide a multitude of PV systems ranging from a few watts to 1–10 kW_p. This may not seem much, but a small outdoor lamp powered by a PV system in Rajasthan, India, for example, has allowed adult literacy classes to be held at night, and projects like this can change the lives of millions. In Kenya, more than 40,000 small PV systems have been installed, financed by a grass-roots private financing program. These projects, and many more not mentioned above, have resulted in increased confidence in, and experience with, PV systems [29].

PV systems are not limited to developing countries. In Dartmount, Devon, UK, for example, a grid-connected roof-top PV system supplies 900 kWh per annum—about 40% of the homes' requirements. In Berlin, at the German nation's leading financial institution, the Berlin Bank, a 50-kW_p system is being tested. In Amersfoort, Netherlands, a 180-KW_p system is being used in a 500 home development. A 4.2-kW_p PV tracking system is installed in Geest in Northern Germany, and in Karlsruhe, the Art and Media Technology building has a PV system that produces 90,000 kWh per year for a DC Tram (transportation) system. In Lausanne, Switzerland, as part of the European Heliotram project, 32 PV modules generate more than 7,800 W_p, which is fed into the DC line of the trolley-bus line. Near the Swiss Alps in Chur, grid-connected PV panels have even been installed on the sound barriers on the motorway N13. Although there are thousands of PV systems in the U.S., one clearly illustrates future trends. In Santa Cruz, California, 224 photovoltaic panels (approx. 12 kW) have been installed at the city hall. A state-sponsored subsidy paid approximately half of the cost of the system. Thanks to the rebates and grants, this PV system, which provides about 7% of the building's power, will pay for itself in fewer than 4 years. These subsidies were needed to make the system attractive compared to conventional energy sources, which are themselves subsidized by government expenditures.

Paradoxically, there is often much talk at the local and national levels about lowering environmental standards so that conventional generators can lower operating costs and pass these saving on to customers. This represents further subsidies to fossil fuels, and distorts the markets away from investments in renewable energy and energy efficiency. In fact, it is the promotion of renewable energy, energy efficiency, and the continuation of current environmental regulations that can benefit the economy in the end. For example, Myers and Kent describe that the U.S. Clean Air Act has produced net direct monetary savings during the period 1970–1990 averaging \$1.1 trillion per year. This is due to the cost of health care associated with pollution and the loss of productivity of people and crops. These are called "external costs," because if the environmental regulations were less restrictive, these costs would be paid by the consumer or society rather than by the energy producer.

In addition, the Kyoto Protocol and related international agreements attempt to limit carbon dioxide emitted during fossil-fuel energy utilization (i.e., coal, oil, natural gas). The concern is that greenhouse gases such as CO_2 will change the balance between incoming solar and outgoing thermal radiation from Earth, and thus the temperature of the planet will be raised. This could lead to climate changes that are difficult to predict. As greenhouse gases are increasingly viewed as a pollutant, there will undoubtedly be increased interest and investment in renewable energy. A popular myth to be dispelled is that c-Si PV modules generate more CO_2 than they offset. Although they do generate CO_2 via the fossil fuels used to make them, they generate approximately 3 times less per unit energy than fossil fuel sources over their 20–30-year lifetime. [25, 28] Changes in energy need not come at the expense of promoting economic growth and development. A Union of Concerned Scientists report asserted that actions to curb global warming are feasible and affordable. Many studies assert that there are many other benefits from renewable energy technologies, including cleaner air and water, the creation of jobs, and the promotion of new technologies and businesses. A switch to renewable energy may also be necessary from an ethical standpoint.

The United States possesses only 4% of the world's population, but it consumes 25% of its energy. The U.S. consumes roughly twice as much energy per person and per unit of GNP as do Western Europeans and the Japanese. By increasing the efficiency with which Americans use energy so that it is equivalent to Western European and Japanese levels, the country could save over $100 billion per year. The U.S. also emits 22% of carbon dioxide accumulating in the global atmosphere (a global common resource). In per-capita terms, it emits twice as much carbon dioxide as Germany, Russia, or Japan, almost three times as much as Italy, and ten times as much as China. Fossil fuels contribute 90% of the United States' greenhouse gas emissions. They also account for 90% of local air pollution and acid rain, and the great majority of gases leading to smog. A United Nations Environment Program study shows that not taking action to curb carbon emissions could lead to more than $300 billion (worldwide) in annual reductions in Gross Domestic Product (GDP) in the future. Many cite the "precautionary principle," which states that in the absence of full understanding of the scientific evidence for an effect with dire, unpredictable, and far-reaching consequences, it is best to err on the side of caution.

While the debate continues over whether the switch to renewable energy is even feasible or possible, renewable energy systems are becoming more common worldwide. As in the skyscraper powered by renewable energy in Times Square, the promise of solar cells is reaching high. Nonpolluting renewable energy systems can continue to make a significant contribution to solving local and global problems. It is up to us to closely examine, and choose, those technologies, economic incentives, national and international policies, and local actions that will hasten the inevitable transition away from fossil fuels toward renewable and sustainable technologies. Solar cells and photovoltaic technologies can, and will, make an impact on the world's energy production. By understanding the basic operation of these remarkable devices, the impact can be made to occur sooner rather than later.

7.5 Conclusions and Further Study

This book has outlined a basic approach to understanding the optical basis of solar cells. The optical properties of solar cell materials, and how these properties relate to the creation of practical devices, have been examined. A simplified version of a generalized model has been outlined and can be used to understand existing quantum-solar-energy converters, as well as solar cells that will be developed in the future. With the completion of this text, the student is ready to go on to more detailed work on the subject. It is hoped that students will do so using the bibliography as a starting point. In many developing nations, and in remote (off-grid) locations, solar cells already provide a highly valued, low-maintenance and cost-effective alternative to fossil fuels and nuclear power. The challenge of the 21st century will be to use optical, electrical, and economic analyses of solar cells to effectively provide energy for a growing population.

Appendix

Table A.1 Standard (ASTM E 892) terrestrial solar spectral irradiance at air mass 1.5 (1000 W/m^2) for a 37-deg tilted surface. [Data courtesy of Keith Emery, NREL, Golden, Colorado. This data is displayed in Fig. 2.2(a).]

λ(μm)	W/m^2/μm	λ(μm)	W/m^2/μm	λ(μm)	W/m^2/μm
0.305	9.5	0.740	1211.2	1.520	262.6
0.310	42.3	0.753	1193.9	1.539	274.2
0.315	107.8	0.758	1175.5	1.558	275.0
0.320	181.0	0.763	0643.1	1.578	244.6
0.325	246.8	0.768	1030.7	1.592	247.4
0.330	395.3	0.780	1131.1	1.610	228.7
0.335	390.1	0.800	1081.6	1.630	244.5
0.340	435.3	0.816	849.2	1.646	234.8
0.345	438.9	0.824	785.0	1.678	220.5
0.350	483.7	0.832	916.4	1.740	171.5
0.360	520.3	0.840	959.9	1.800	30.7
0.370	666.2	0.860	978.9	1.860	2.0
0.380	712.5	0.880	933.2	1.920	1.2
0.390	720.7	0.905	748.5	1.960	21.2
0.400	1013.1	0.915	667.5	1.985	91.1
0.410	1158.2	0.925	690.3	2.005	26.8
0.420	1184.0	0.930	403.6	2.035	99.5
0.430	1071.9	0.937	258.3	2.065	60.4
0.440	1302.0	0.948	313.6	2.100	89.1
0.450	1526.0	0.965	526.8	2.148	82.2
0.460	1599.6	0.980	646.4	2.198	71.5
0.470	1581.0	0.993	746.8	2.270	70.2
0.480	1628.3	1.040	690.5	2.360	62.0
0.490	1539.2	1.070	637.5	2.450	21.2
0.500	1548.7	1.100	412.6	2.494	18.5
0.510	1586.5	1.120	108.9	2.537	3.2
0.520	1484.9	1.130	189.1	2.941	4.4
0.530	1572.4	1.137	132.2	2.973	7.6
0.540	1550.7	1.161	339.0	3.005	6.5
0.550	1561.5	1.180	460.0	3.056	3.2
0.570	1501.5	1.200	423.6	3.132	5.4

Table A.1 Continued.

λ(μm)	$W/m^2/\mu m$	λ(μm)	$W/m^2/\mu m$	λ(μm)	$W/m^2/\mu m$
0.590	1395.5	1.235	480.5	3.156	19.4
0.610	1485.3	1.290	413.1	3.204	1.3
0.630	1434.1	1.320	250.2	3.245	3.2
0.650	1419.9	1.350	32.5	3.317	13.1
0.670	1392.3	1.395	1.6	3.344	3.2
0.690	1130.0	1.443	55.7	3.450	13.3
0.710	1316.7	1.463	105.1	3.573	11.9
0.718	1010.3	1.477	105.5	3.765	9.8
0.724	1043.2	1.497	182.1	4.045	7.5

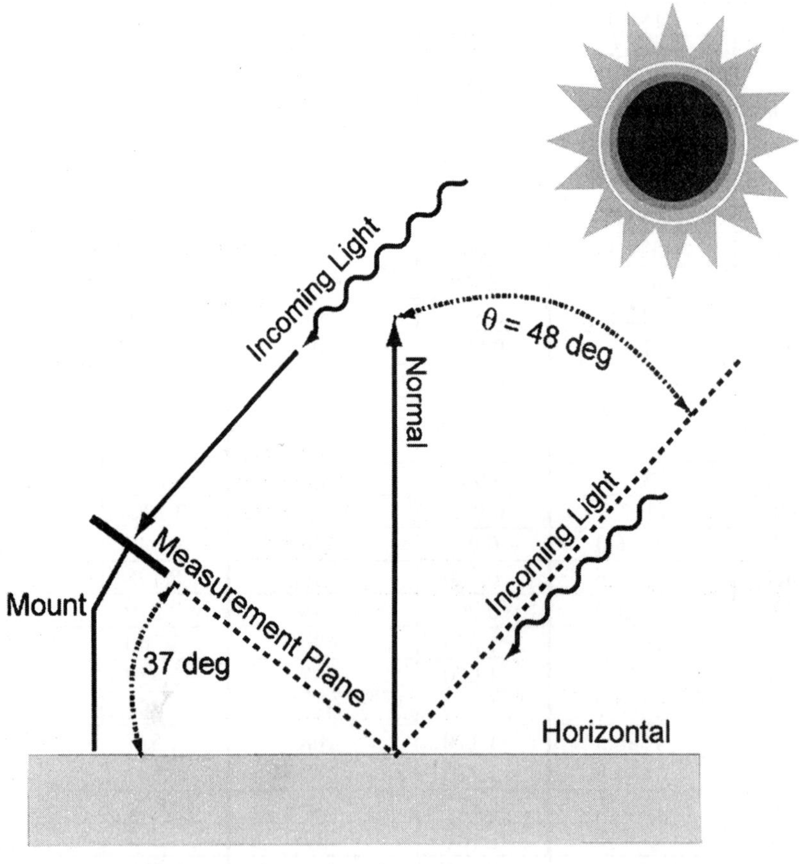

Fig. A.1 The geometry that defines the standard for the terrestrial solar spectral irradiance tables at air mass 1.5 for a 37-deg tilted surface.

A Basic Solar Energy Library for the Optical Specialist

The following is a partial list of the recommended references in the field of solar energy conversion and photovoltaics. These books and other sources can be explored for details on the topics introduced in this brief tutorial.

1. M. A. Green, *Silicon Solar Cells: Advanced Principles and Practice*, Bridge Printery, Sydney, 1995. This book (and any of Green's many articles found in the literature) is a must-read. It also contains useful constants and tables for silicon and a description of texturing the front of a solar cell.

2. American Society for Testing and Materials, *Standard for Terrestrial Solar Spectral Irradiance Tables at Air Mass 1.5 for a 37° Tilted Surface*, ASTM standard E 892, Vol. 12.02, West Conshohocken, Penn., 2002 Annual Book of ASTM Standards. This standard is equivalent to IEC 60904-3 and ISO 9845-1. ASTM standards E 948 and E 1021 are also useful for measurement of the I-V curve and Spectral Response curve, respectively.

3. H. J. Möller, *Semiconductors for Solar Cells*, Artech House, Boston, 1993. Covers the materials science aspects of solar cells and gives specific examples of technology applications.

4. T. Markvart, *Solar Electricity*, 2nd Edition, John Wiley & Sons, New York, 2000. A more recent work exploring all aspects of solar cells.

5. A. L. Fahrenbrunch, R. H. Bube, *Fundamentals of Solar Cells*, Academic Press, 1983. One of the most comprehensive books on solar cell modeling equations.

6. A. Goetzberger, J. Knobloch, B. Voss, *Crystalline Silicon Solar Cells*, John Wiley & Sons, New York, 1998. This book covers solar cell device equations and characterization techniques. It is written by well-known authors from one of the leading solar laboratories in the world, the Fraunhofer Institute in Freiburg, Germany.

7. J. I. Pankove, *Optical Processes in Semiconductors*, Dover, New York, 1971. This is the classic book that describes the optical aspects of materials and it makes worthwhile reading for anyone in the optical field.

8. *Encyclopedia of Electrochemistry*, Volume 6, *Semiconductor Electrodes and Photoelectrochemistry*, John Wiley & Sons, New York, 2001. This work gives details on the dye-sensitized solar cell. The references therein may be useful for those who want details on photoelectrochemical solar cells.

9. C. Winter, R. Sizman and L. Vant Hull, *Solar Power Plants*, Springer-Verlag, New York, 1991. Details the thermodynamic and practical aspects of solar cells, concentrators, and systems.

10. J. Duffie and W. Beckman, *Solar Engineering of Thermal Processes* 2nd Edition, John Wiley & Sons, Inc., New York, 1991, pp. 926–931. Covers material on solar energy (thermal and PV) and solar radiation detection.

11. G. Smestad, P. Hamill, "Concentration of solar radiation by white backed photovoltaic panels," *Applied Optics,* **23**, pp. 4394, 1984.

12. G. Smestad, H. Ries, R. Winston, E. Yablonovitch, "Thermodynamic limits of light concentrators," *Solar Energy Materials and Solar Cells,* **21**, p. 95, 1990.

13. G. Smestad, H. Ries, "Luminescence and current-voltage characteristics of solar cells and optoelectronic devices," *Solar Energy Materials and Solar Cells,* **25**, pp. 51, 1992.

14. G. Smestad et al., "Dye sensitized TiO_2 solar cells I & II," *Solar Energy Materials and Solar Cells,* **32**, pp. 259, 1994.

15. G. Smestad, "Absorptivity as a predictor of the photoluminescence spectra of silicon solar cells and photosynthesis," *Solar Energy Materials and Solar Cells,* **38**, pp. 57, 1995.

16. G. Smestad, "Demonstrating electron transfer and nanotechnology: a natural dye-sensitized nanocrystalline energy converter," *Journal of Chemical Education,* **75**, pp. 752, 1998. Papers 11–16 may prove helpful in building a more detailed understanding of the aspects of solar cells covered in this tutorial.

17. B. Andresen, R. S. Berry, M. Ondrechen and P. Salamon, *Accounts of Chemical Res.,* **17**, pp. 266, 1984.

18 B. Andresen, P. Salamon and R. S. Berry, *Physics Today,* September, 1984. These articles discuss the general thermodynamics of energy conversion.

19. A. De Vos, *Endoreversible Thermodynamics of Solar Energy Conversion*, Oxford Univ. Press, Oxford, 1992. A mathematical treatment of the fundamental (thermodynamic) aspects of all solar converters. It covers thermal and photo-conversion, photosynthesis, and tandem systems; and it provides useful insights for those who wish to determine the limitations of a given approach.

20. A. Luque, G. L. Araujo, *Physical Limitations to Photovoltaic Energy Conversion*, Adam Hilger, New York, 1990. Researchers at the leading lab in Spain cover thermodynamic aspects of solar conversion. The book has a good section on light concentrators.

21. M. Andreev, V. A. Grilikhes, V. D. Rumyantsev, *Photovoltaic Conversion of Concentrated Sunlight*, John Wiley & Sons, New York, 1997. This book offers a look at materials science aspects of solar cells as used in concentrators. These Russian authors bring together a wealth of information on solar cells from the long history of research in the former Soviet Union.

22. P. Würfel, S. Finkbeiner, E. Daub, "Generalized Planck's radiation law for luminescence via indirect transistions," *Appl. Phys. A,* **60**, pp. 67, 1995.

23. W. T. Welford, R. Winston, *High Collection Non-Imaging Optics*, Academic Press, New York, 1989. This is a good source for optical concentrators from the researchers who invented the CPC.

24. D. Pimentel, G. Rodriguez, "Renewable energy: economic and environmental issues," *Bio. Sci.,* **44**, pp. 536, 1994.

25. J. Turner, "A realizable renewable energy future," *Science,* **285**, p. 687. References 24–25 are good sources for the potential for solar and other renewable energy.

26. E. Martinot, "Renewable energy investment by the World Bank," *Energy Policy,* **29,** p. 689, 2001.

27. N. Myers and J. Kent, *Perverse Subsidies: Tax Dollars Undercutting Our Economies and Environments Alike*, International Institute for Sustainable Development, and Island Press, Covelo, CA, 1998. Covers both energy and transportation, as well as other subsidies.

28. M. Oliver, T. Jackson, "The evolution of economic and environmental cost for crystalline silicon photovoltaics," *Energy Policy,* **28,** pp. 1011, 2000.

29. Useful websites:
National Renewable Energy Laboratory, http://www.nrel.gov/
Renewable Energy Policy Project, http://www.repp.org/

Index

Greg P. Smestad received his Ph.D. in Physical Chemistry from the Swiss Federal Institute of Technology in Lausanne under the guidance of Drs. Michael Grätzel and Harald Ries. He received his M.S. in Materials Science in 1985 from Stanford University and his B.S. in Biology in 1983 from the University of Santa Clara, California. He currently works in the optoelectronics industry.

From 1985 to 1990, Dr. Smestad was employed by the Hewlett-Packard Company, first as a materials engineer, and later as an optics design engineer. From 1990 to 1992, he created and managed the optics lab at the Hahn-Meitner Institute Solar Energy group in Berlin, Germany, and conducted research on iron pyrite photoelectrochemical solar cells. In 1992, he became part of the Solar Chemistry group of the Energy and Process Technology department at the Paul Scherrer Institute in Villigen, Switzerland. In 1994, he was employed by Lawrence Berkeley National Laboratory as an electrochemist and researcher on the topic of electrochromic windows. From 1995 to 2002, he was a professor, first as a chemist at the California State University, Monterey Bay, and then as an energy science-policy specialist at the Monterey Institute of International Studies. From 1996 to 1999, he served as a consultant for Prof. Grätzel at the Swiss Federal Institute of Technology, and from 1990 onward he has served as Assistant Editor for the Elsevier journal *Solar Energy Materials and Solar Cells*.

Dr. Smestad has also served as a conference chair at international conferences, authored more than 15 scientific papers in technology and materials science, and holds three U.S. patents. He enjoys combining his interests to solve practical problems in chemistry, solar and renewable energy, and energy policy.